Heat Analysis and Thermodynamic Effects

Heat Analysis and Thermodynamic Effects

Editor

Petros Antonis

scitus
academics

Heat Analysis and Thermodynamic Effects
Edited by **Petros Antonis**

ISBN: 978-1-68117-310-8
Library of Congress Control Number: 2016937619

SCITUS Academics LLC,
www.scitusacademics.com
Box No. 4766, 616 Corporate Way,
Suite 2, Valley Cottage,
NY 10989

Notice

Printed in United States of America on Acid Free Paper

Preface

Heat is defined in physics as the transfer of thermal energy across a well-defined boundary around a thermodynamic system. The transfer of heat is normally from a high temperature object to a lower temperature object. Heat transfer changes the internal energy of both systems involved according to the First Law of Thermodynamics. Heat transfer is the exchange of thermal energy between physical systems. The rate of heat transfer is dependent on the temperatures of the systems and the properties of the intervening medium through which the heat is transferred. The direction of heat transfer is from a region of high temperature to another region of lower temperature, and is governed by the Second Law of Thermodynamics. Heat transfer changes the internal energy of the systems from which and to which the energy is transferred. Heat transfer will occur in a direction that increases the entropy of the collection of systems. The heat transfer and analysis on heat pipe and exchanger, and thermal stress are significant issues in a design of wide range of industrial processes and devices. This book entitled Heat Analysis and Thermodynamic Effects covers mainly thermodynamic effects and thermal stress, heat pipe and exchanger, gas flow and oxidation, and heat analysis. The advanced concepts and information described in this book will be fruitful for the readers to find a sustainable solution in an industrialized society.

Table of Contents

1

CHAPTER

ANALYSIS OF THERMODYNAMIC CHARACTERISTIC CHANGES IN DIRECT EXPANSION GROUND SOURCE HEAT PUMP USING HYDROFLUOROOLEFINS (HFOS) SUBSTITUTING FOR HFC-134A

Yuefen Gao[1,2], Honglei Zhao[1], Yingxin Peng[1], Tony Roskilly[2]

[1]*School of Energy, Power & Mechanical Engineering, North China Electric Power University, Baoding, China*

[2]*Swan Center for Energy Research, Newcastle University, Newcastle Upon Tyne, UK*

ABSTRACT

HFO-1234yf and HFO-1234ze[E] have low global warming potential and zero ozone depletion potential. If they are used in the direct expansion ground source heat pump system substituting for HFC-134a, the system will be beneficial to mitigating climate change. This study aims to find out the thermodynamic characteristics of the direct expansion ground source heat pump system using HFO-1234yf or HFO-1234ze[E] by theoretical calculation. The results indicate that HFO-1234yf system in an actual cycle has the highest COP. HFO-1234yf and HFO-1234ze[E] have such smaller capacity per unit of swept volume that they need larger compression capacity if providing the same heating or cooling loads. For a given unit when HFC-134a is replaced with HFO-1234yf or HFO-1234ze[E], the capacity will decrease. More refrigerant charge is required in the

HFO-1234yf or HFO-1234ze[E] system. The results also present that more refrigerant charge is required in the cooling mode than in the heating mode.

Keywords: Direct Expansion Ground Source Heat Pump (DXGSHP); Thermodynamic Performance; HFO-1234yf; HFO-1234ze[E]; HFC-134a

1. INTRODUCTION

Direct expansion ground source heat pump (DXGSHP) has only copper loops circulating refrigerants which exchanges heat directly with the soil through the walls of the copper tubing. It is an energy-efficient and environmentally clean space conditioning system. However, it has the risk of ground contamination if refrigerant leak into the ground, as the loops containing refrigerant are directly buried in the ground.

One of the solutions to this problem is to select non-toxic, environment-amiable working fluids. Generally, commercial and residential heat pump systems use hydrofluorocarbons (HFCs) or their mixtures as refrigerants. HFCs have zero ozone depletion potential (ODP), but most of them have relatively high values of global warming potentials (GWP). In developed countries, refrigerants with high GWP are facing to be phased out. The European Union's F-gas regulation and the directive 2006/40/EC ban fluorinated gases having GWP greater than 150 in new mobile models from January 1, 2011 and in new vehicles from January 1, 2017 [1,2]. More efforts have been underway to investigate fluorinated propene isomers as possible refrigerants. In 2007, DuPont and Honeywell co-developed hydrofluoroolefins (HFOs) to replace HFCs in air conditioning units. HFOs are a class of compounds. Among the compounds, HFO-1234yf and HFO-1234ze[E] are the two most suitable to the air conditioning system [3].

HFO-1234yf and HFO-1234ze[E] possess the similar thermodynamic behavior to HFC-134a. They have a 100 year GWP of 4 and 6 respectively. Both of them are being considered as a possible replacement for HFC-134a which has a high GWP value of 1430 [3]. The US EPA has released a proposed rule for HFO-1234yf as an automotive refrigerant [4].

Not only are HFOs able to be used to the air conditioning system in vehicles, they are also feasible to replace HFC-134a in commercial and residential heat pump system. Many investigations released that HFO-1234yf and HFO-1234ze[E] almost have non-toxic and environmental impact [3,5-10]. They are also compatible with the lubricate oil and materials normally used in HFC-134a system [5,11]. HFO-1234ze[E] is non-flammable, while HFO-1234yf is quite mildly flammable [3]. So they are ideal options for DXGSHP in terms of climate change and safety.

Some research results also indicated that HFOs' thermophysical property parameters are to some extent different from HFC-134a's [3,5,12-15]. Correspondingly, the DXGSHP system possibly has different thermodynamic performance with these refrigerants.

This study aims to compare the thermodynamic performances of the DXGSHP system using the three kinds of refrigerants and to recommend the suitable alternative substance for HFC-134a based on theoretical calculation and analysis.

2. DXGSHP SYSTEM DESCRIPTION

In the DXGSHP system, the refrigerant loops are directly buried underground and exchange heat with the ground.

Without using an intermediate fluid, the DXGSHP system is significantly more efficient than the conventional GSHP system. Above all, the phase change of the refrigerant occurs in the ground, where latent heat is efficiently rejected or absorbed. The elimination of the water pump and the water heat exchanger also greatly reduce electricity consumption and heat losses.

Figure 1 shows a schematic diagram of DXGSHP system. The system operates in both heating and cooling mode. A reversal valve transforms the operating mode.

In the heating mode, refrigerant absorbs heat from the ground through the loops named ground heat exchanger (GHX). It then enters the compressor where it is compressed to high-temperature vapor. The discharged vapor then enters air heat exchanger (AHX) where it releases heat to the space and condenses to liquid. The liquid exits from the AHX and is throttled via the expansion valve (EV). It then enters the GHX.

In the cooling mode, inversely, the refrigerant releases heat to the ground via GHX. The refrigerant absorbs heat from the space via AHX. It then enters the compressor and is compressed. The discharged vapor enters GHX and condenses to liquid. The liquid exits from the GHX and is throttled via the EV, and finally enters the AHX.

The thermodynamic cycle is expressed in the p-h diagram (Figure 2). Both the modes have the same cycle diagram except for the parameters.

3. SIMULATING CALCULATION OF THE THERMODYNAMIC PERFORMANCE OF DXGSHP

In order to compare the thermodynamic performance of the DX-GSHP using the three different kinds of refrigerants, simulating calculation are conducted based on the following conditions and assumptions. In the thermodynamic calculations, the thermophysical property parame ters of each state point are based on the program REF PROP8.0 for HFC-134a, Extended Corresponding States (ECS) model for HFO-1234ze(E) [16] and Martin-Hou Equation of State (EOS) for HFO-1234yf [5]. The ECS model has adequate accuracy [16,17].

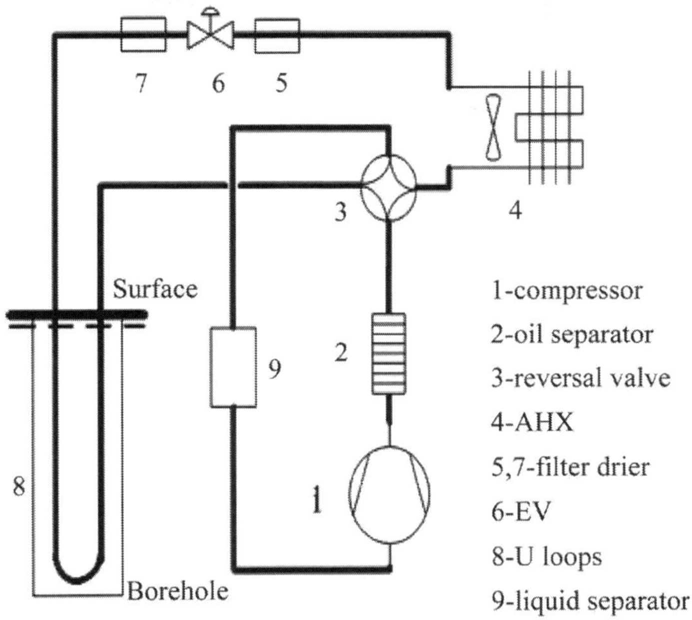

Figure 1.The schematic of the DX-GCHP system.

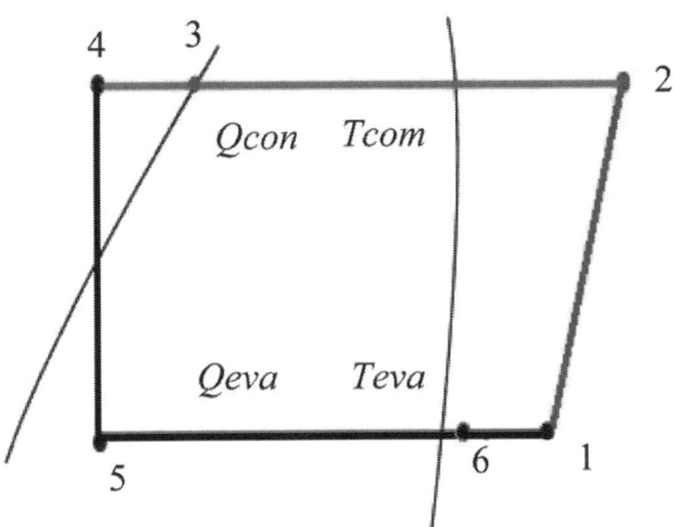

Figure 2. The p-h diagram of thermodynamic cycle.

3.1. Parameters at Design Conditions

Evaporating temperature and condensing temperature are the two most important parameters in the system design. The temperatures depend much on the external parameters. According to ANSI-AHRI870-2005, in the cooling mode, air temperature entering indoor AHX is 26.7°C dry-bulb, 19.4°C wet-bulb. Refrigerant temperature of liquid line from GHX is 25.0°C. In the heating mode, air temperature entering AHX is 21.1°C dry-bulb, 15.6°C wet-bulb. Refrigerant temperature of vapor line from GHX is 0°C [18]. The values of evaporating temperature and condensing temperature are presented in Table 1.

3.2. Assumptions in the Thermodynamic Calculation

- All the processes are under steady state conditions.
- There are no potential or kinetic energy effects and no chemical or nuclear reactions.
- Heat losses and refrigerant pressure drops in the connecting tubes are negligible.

Table 1. Parameters at design conditions.

Parameters	Cooling mode	Heating mode
Air temperature entering AHX	26.7°C dry-bulb 19.4°C wet-bulb	21.1°C dry-bulb 15.6°C wet-bulb
Evaporating temperature, teva	7°C	0°C
Superheat degree, Δtsh	5°C	5°C
Suction temperature, tsuc	17°C	10°C
Condensing temperature, tcon	30°C	35°C
Subcooling degree, Δtsc	5°C	5°C

- The isentropic efficiency of the compressor is 85%. The compressor mechanical efficiency and the compressor motor electrical efficiency are 70% and 75%, respectively.
- The systems with different refrigerants have the same compressing speed.

3.3. Models Assessing the Thermodynamic Performance of DXGSHP System

The performance of the system is evaluated primarily in terms of the capacity and coefficient of performance (COP). The parameters are calculated according to the steady flow energy equations.

The capacity is related to the operation parameters and the refrigerant mass flow rate. The relations are expressed as

$$Q_{eva} = m_r \left(h_{eva,o} - h_{eva,i} \right) \tag{1}$$

$$Q_{con} = m_r \left(h_{con,i} - h_{con,o} \right) \tag{2}$$

$$m_r = \eta_v \frac{V_h}{vs_{uc}} \tag{3}$$

$$\eta_v = A - B \left(\frac{P_{dis}}{P_{suc}} \right)^{\frac{1}{n}} \tag{4}$$

where, η_v is the volumetric efficiency, A and B are the coefficients related to the types of the compressor, n is the Specific heat ratio related to the refrigerant type.

The power input is related to the operation parameters, the mass flow rate, the mechanical efficiency of the compressor and the electrical efficiency of the driving motor.

$$W_{com} = m_r \frac{h_{dis} - h_{suc}}{\eta_m \eta_e} \quad (5)$$

η_m tends to decrease with the increase of pressure ratio.

In the heating mode COP_h is the ratio of the heating capacity Q_{con} to the power input W_{com}, whereas, in the cooling mode COP_c is the ratio of the cooling capacity $Q_{eva,m}$ to the power input W_{com}, as given below respectively.

$$COP_h = \frac{Q_{con}}{W_{com}} \quad (6)$$

$$COP_c = \frac{Q_{eva}}{W_{com}} \quad (7)$$

Combining the equations of (1), (5), (7) and (2), (5), (6), COP_h and COP_c can be expressed as

$$COP_h = \eta_m \eta_e \frac{h_{con,i} - h_{con,o}}{h_{dis} - h_{suc}} \quad (8)$$

$$COP_c = \eta_m \eta_e \frac{h_{eva,o} - h_{eva,i}}{h_{dis} - h_{suc}} \quad (9)$$

4. RESULTS AND ANALYSIS

4.1. Calculating Results

Based on the parameters in Table 1 and the assumptions above, the calculating results are shown in **Tables 2-4**.

Table 2 states the typical characteristic parameters and performance values. In the theoretical cycle, HFC-134a has the highest capacity per unit mass flow rate. HFO-1234ze[E] has the closest values to HFC-134a. However, HFO-1234yf consumes the lowest power among them. The ratio of $COP_{th,h}$ of the three kinds of refrigerants (HCF-134a, HFO-1234yf, HFO-1234ze[E]) is 1:0.995: 1.041, while the ratio of $COP_{th,c}$ of the three kinds of refrigerants is 1:1.085:1.141. That means HFO-1234ze[E] has the best thermodynamic performance both in the heating mode and in the cooling mode, while HFO- 1234yf has the lowest performance in the heating mode. But, in the actual cycle, the ratio of COP_h of the three kinds of refrigerants (HCF-134a, HFO-1234yf, HFO-1234ze[E]) is 1:1.028:0.955, while the ratio of COP_c is 1:1.121:1.008. HFO-1234yf has the best thermodynamic performance.

Table 3 shows the calculating values and the ratios under the conditions that all refrigerants have the same cooling capacity and the mass flow rate of each refrigerant itself keeps the same in each operating mode. The system using HFO-1234yf needs 23% more mass charge than the system using HFC-134a, while the mass charge of the system using HFO-1234ze[E] is nearly 10% more than that of the system using HFC-134a. Under the same condition, the volume flow rate of HFO-1234ze[E] system is 24% - 26% more than that of HFC-134a system.

Table 4 gives the results that all refrigerants have the same volume flow rate in each operating mode. Refrigerant charge is apt for the maximum in each operating mode. The mass flow rate is regulated according to the operating mode. A reservoir is installed to contain the redundant refrigerant. The results show that refrigerant charge of HFO-1234yf system is 21% more than that of HFC-134a system, whereas cooling capacity and heating capacity of HFO-1234yf system are 2% and 8% less than those of HFC-134a system respectively. As for HFO-1234ze[E] system, refrigerant charge of the system is 14% less than that of HFC-134a system, whereas heating capacity and cooling capacity of HFO-1234ze[E] system are about 20% less than those of HFC-134a system.

Table 2. Thermodynamic parameters.

Variables	Cooling mode			Heating mode		
	HFC-134a	HFO-1234yf	HFO-1234ze[E]	HFC-134a	HFO-1234yf	HFO-1234ze[E]
v_{suc}, m^3/kg	0.058	0.048	0.067	0.073	0.063	0.083
p_{dis}/p_{suc}	2.03	1.95	2.15	2.98	2.86	3.05
t_{dis}, °C	43	38	40	48	44	44
$q_{eva,v}$, kJ/m^3	2924.8	2864.6	2324.9	2154.4	1989.4	1742.3
$q_{eva,m}$, kJ/kg	169.639	137.500	155.769	157.273	125.334	144.615
q_{con}, kJ/kg	186.363	150.000	169.231	181.818	145.000	166.153
w_{com}, kJ/kg	27.878	20.161	25.400	40.908	31.721	39.160
$COP_{th,h}$	-	-	-	7.407	7.373	7.714
$COP_{th,c}$	10.142	11.000	11.571	-	-	-
COP_h	-	-	-	4.445	4.571	4.243
COP_c	6.085	6.820	6.133	-	-	-

Table 3. Thermodynamic variables based on the same cooling capacity.

Variables	Cooling Mode			Heating Mode		
	HFC-134a	HFO-1234yf	HFO-1234ze[E]	HFC-134a	HFO-1234yf	HFO-1234ze[E]
$V_r \times 10^{-3}$, m^3/s	2.051	2.094	2.581	2.583	2.747	3.195
m_r, kg/s	0.0354	0.0436	0.0385	0.0354	0.0436	0.0385
Q_{eva}, kW	6	6	6	5.563	5.465	5.568
Q_{con}, kW	6.592	6.545	6.518	6.431	6.319	6.395
W_{com}, kW	0.987	0.879	0.978	1.447	1.383	1.508
$m_{r1}:m_{r2}:m_{r3}$		1:1.23:1.09			1:1.23:1.09	
$V_{r1}:V_{r2}:V_{r3}$		1:1.02:1.26			1:1.06:1.24	
$Q_{con1}:Q_{con2}:Q_{con3}$		1:0.993:0.989			1:0.983:0.994	
$Q_{eva1}:Q_{eva2}:Q_{eva3}$		1:1:1			1:0.982:1.001	

Table 4. Thermodynamic variables with the same volume flow rate.

Variables	Cooling Mode			Heating Mode		
	HFC-134a	HFO-1234yf	HFO-1234ze[E]	HFC-134a	HFO-1234yf	HFO-1234ze[E]
$V_r \times 10^{-3}$, m^3/s	2.051	2.051	2.051	2.051	2.051	2.051
m_r, kg/s	0.0354	0.0427	0.0306	0.0281	0.0326	0.247
Q_{eva}, kW	6	5.871	4.773	4.418	4.081	3.574
Q_{con}, kW	6.592	6.404	5.186	5.107	4.721	4.106
W_{com}, kW	0.987	0.861	0.778	1.149	1.034	0.968
$m_{r1}:m_{r2}:m_{r3}$		1:1.21:0.86			1:1.16:0.88	
$V_{r1}:V_{r2}:V_{r3}$		1:1:1			1:1:1	
$Q_{con1}:Q_{con2}:Q_{con3}$		1:0.971:0.787			1:0.924:0.804	
$Q_{eva1}:Q_{eva2}:Q_{eva3}$		1:0.979:0.796			1:0.924:0.809	

4.2. Analyses and Discussions

4.2.1. Thermophysical Properties

1) Pressure ratio As for the refrigerating cycle, the ratio of the discharge pressure p_{dis} to the suction pressure p_{suc} affects the thermodynamic performance significantly. Equations (1)-(9) show that the volumetric efficiency η_v, the mass flow rate m_r, the cooling capacity Q_{eva} or heating capacity Q_{con}, the power consumption W_{com} and COP are the functions of the pressure ratio. The adiabatic indicated efficiency and the friction efficiency decline as the pressure ratio increases.

HFO-1234ze[E] has the highest pressure ratio (p_{dis}/p_{suc}), while HFO-1234yf has the lowest. In Table 2, p_{dis}/p_{suc} of HFO-1234ze[E] and HFO-1234yf are 2.15 and 1.95 respectively in the cooling mode, 3.05 and 2.86 in the heating mode. So, the higher pressure ratio of HFO-1234ze[E] in an actual cycle offsets its original advantages. The ratio of COP_h of the three kinds of refrigerants (HCF-134a, HFO-1234yf, HFO-1234ze[E]) is 1:1.028:0.955, while the ratio of COP_c is 1:1.121:1.008. As discussed above, in the theoretical cycle, HFO-1234ze[E] has the highest value of COP both in the cooling and heating mode. However, in the actual cycle HFO-1234yf obviously has the best thermodynamic performance because of its lower pressure ratio.

2) Working pressure and temperature When the systems are running under the same conditions, the suction pressure and the discharge pressure of HFO-1234yf are slightly higher than those of HCF-134a. However, HFO-1234ze[E] is just the reverse. The discharge vapor temperature of both HFO-1234yf and HFO-1234ze[E] is lower than HCF-134a. So when HFO-1234yf or HFO-1234ze[E] directly substitutes for HFC-134a, the compressor or the system still works in safety, as the maximum working pressure and temperature are kept below or close to the values of the HFC-134a system.

3) Suction specific volume v_{suc}

Suction specific volume v_{suc} affects the volume flow, and further the size of the compression chamber, evaporator, condenser and pipes.

HFO-1234ze[E] has the largest specific volumes (0.067 m^3 /kg in the cooling mode, 0.083 m^3/kg in the heating mode), while HFO-1234yf has the smallest ones (0.048 m^3/kg in the cooling mode, 0.063 m^3/kg in the heating mode). So, the capacity per unit of swept volume in the HFO-1234ze[E] system is the smallest. Larger size compression chamber, evaporator, condenser and pipes are needed. This will raise the initial investment.

4.2.2. Refrigerant Charge

1) Refrigerant mass charge varying in different systems Two cases are discussed here. The first case is that the three systems provide the same cooling capacity. The second case is that the volume flow rates in the systems keep at a given value.

In the first case, more refrigerant charge is required in the HFO-1234yf and HFO-1234ze[E] systems when they provide the same cooling capacity. The data in Table 3 show that the system using HFO-1234yf needs 23% more mass charge than the system using HFC-134a, while the mass charge of the system using HFO-1234ze[E] is nearly 10% more than that of the system using HFC-134a.

According to the results in Table 3, the volume flow rate of HFO-1234ze[E] system is 24-26% more than that of HFC-134a system. So the compression chamber and the pipes of HFO-1234ze[E] system should have larger sizes. The capacities of condenser in the heating mode (i.e., the heating capacities) of HFO-1234yf and HFO-1234ze[E] system are 2% and 1% less than that of HFC-134a system respectively.

In the second case, when HFC-134a is replaced with HFO-1234yf or HFO-1234ze[E] in a given unit, the heating capacity and the cooling capacity will decrease.

The results in Table 4 show that refrigerant charge of HFO-1234yf system is 21% more than that of HFC-134a system, whereas cooling capacity and heating capacity of HFO-1234yf system are 2% and 8% less than those of HFC-134a system respectively. As for HFO-1234ze[E] system, refrigerant charge of the system is 14% less than that of HFC-134a system, whereas heating capacity and cooling capacity of HFO-1234ze[E] system are about 20% less than those of HFC-134a system.

2) More refrigerant charge required in the cooling mode.

More refrigerant charge is required in the cooling mode than in the heating mode.

The results in Table 4 show that 24% - 31% more refrigerant charge is required in the cooling mode than in the heating mode.

4.2.3. Compression Chamber Volume

HFO-1234yf and HFO-1234ze[E] need larger compression chamber volume if they provide the same heating or cooling loads, because HFO-1234yf and HFO-1234ze[E] have the smaller capacity per unit of swept volume $q_{eva,v}$ than HFC-134a. According to Table 2, $q_{eva,v}$ of HFC-134a, HFO-1234yf and HFO-1234ze[E] systems are 2924.8 kJ/m^3, 2864.6 kJ/m^3, 2324.9 kJ/m^3 respectively in the cooling mode, and 2154.4 kJ/m^3, 1989.4 kJ/m^3, 1742.3 kJ/m^3 in the heating mode.

4.2.4. Capacity and COP

Capacity and COP characterize the capability and performance of the system.

As for the cooling or heating capacity, HFC-134a system has the largest capacity per unit of swept volume $q_{eva,v}$. HFO-1234yf system has the similar value to HFC-134a system. But the capacity of HFO-1234ze[E] system is about 20% less than that of HFC-134a system. That means HFO-1234yf is the better replacement for HFC-134a than HFO-1234ze[E].

In terms of COP, HFO-1234yf has the better performance than the other two in the actual cycle, as discussed above.

5. CONCLUSIONS

In the DXGSHP system, the refrigerant loops are directly buried in the ground, which is in risk of ground contamination. Using non-toxic refrigerants is one effective way to solve this problem. Both HFO-1234yf and HFO-1234ze[E] are ideal potential substituent for HFC-134a in terms of climate change and safety.

HFO-1234ze[E] has the best thermodynamic performance assumed that all the refrigerants have the same mass flow rate. However, in an actual cycle HFO-1234yf has the best thermodynamic performance due to its low pressure ratio.

More refrigerant mass charges are required in the HFO-1234yf and HFO-1234ze[E] system when keeping the heating capacity the same. Moreover, more refrigerant mass charge is required in the cooling mode than in the heating mode.

HFO-1234yf and HFO-1234ze[E] have so smaller capacity per unit of swept volume that they need larger chamber if providing the same heating or cooling loads. For a given unit when HFC-134a is replaced with HFO-1234yf or HFO-1234ze[E], the capacity will decrease.

In conclusion, HFO-1234yf is the better potential alternative substance for HFC-134a. HFO-1234ze[E] can also substitute for HFC-134a, but it needs larger size unit.

6. ACKNOWLEDGEMENTS

Yuefen Gao, Honglei Zhao and Yingxin Peng are grateful to the financial support by the Fundamental Research Funds for the Central Universities (Project No: 11MG43). This work is also partly supported by European Regional Development Fund (ERDF 2009-12).

REFERENCES

1. The European Parliament and the Council of the European Union, "Regulations (EC) No 842/2006 of the European Parliament and of the Council of 17 May 2006 on Certain Fluorinated Greenhouse Gases," Official Journal of the European Union, 2006. http://eur-lex.europa.eu /LexUriServ/LexUriServ.do?uri=OJ:L:2006:161:0001:0011:E N:PDF

2. The European Parliament and the Council of the European Union, "Directive 2006/40/EC of the European Parliament and of the Council of 17 May 2006 Relating to Emissions from Air-Conditioning Systems in Motor Vehicles and Amending Council Directive 70/156/EEC," Official Journal of the European Union, 2006. http://eur-lex.europa.eu/ LexUriServ/LexUriServ.do?uri=OJ:L:2006:161:0012:0018:EN :PDF

3. T. G. A. Vink, "Synthetic and Natural Refrigeration Fluids Recent Developments," 2012. http://www.fluorocarbons.org /uploads/Modules/Library/tvink_honeywell_ecocool10.pdf

4. USEPA, "Protection of Stratospheric Ozone: New Substitute in the Motor Vehicle Air Conditioning Sector Under the Significant New Alternatives Policy (SNAP) Program," Federal Register, Vol. 74, No. 200, 2009. http://www.gpo.gov/fdsys/pkg/FR-2009-10-19/pdf/E9-25106.pdf

5. T. J. Leck, "Evaluation of HFO-1234yf as a Potential Replacement for R-134a in Refrigeration Applications," Third IIR Conference on Thermophysical Properties and Transfer Processes of Refrigerants, Boulder, 23-26 June 2009. http://www2.dupont.com/Refrigerants/en.../HFO-1234yf_IIR_Leck.pdf

6. C. Zilio, J. S. Brown, G. Schiochet and A. Cavallini, "The Refrigerant R1234yf in Air Conditioning Systems," Energy, 36, No. 10, 2011, pp. 6110-6120.

7. P. Schuster, R. Bertermann, G. M. Rusch and W. Dekant, "Biotransformation of trans-1,1,1,3-tetrafluoropropene (HFO-1234ze)," Toxicology and Applied Pharmacology Vol. 239, No. 3, 2009, pp. 215-223.

8. P. Schuster, R. Bertermann, T. A. Snow, X. Han, G. M. Rusch and G. W. Jepson, "Biotransformation of 2,3,3,3-tetrafluoropropene (HFO-1234yf)," Toxicology and Applied Pharmacology, Vol. 233, No. 3, 2008, pp. 323-332.

9. T. J. Wallington, M. P. S. Andersen and O. J. Nielsen, "Estimated Photochemical Ozone Creation Potentials (POCPs) of CF_3CFCH_2 (HFO-1234yf) and Related Hydrofluoroolefins (HFOs)," Atmospheric Environment, Vol. 44, No. 11, 2010, pp. 1478-1481.

10. K. Takizawa, K. Tokuhashi and S. Kondo, "Flammability Assessment of CH_2CFCF_3: Comparison with Fluoroalkenes and Fluoroalkanes," Journal of Hazardous Materials, Vol. 172, No. 2-3, 2009, pp. 1329-1338.

11. B. Wang, W. Zhang and J. Lv, "A New Refrigerant HFO-1234ze," New Chemical Materials, Vol. 36, No. 2, 2008, pp. 10-12. (in Chinese)

12. K. Tanaka and Y. Higashi, "Thermodynamic Properties of HFO-1234yf (2,3,3,3-tetrafluoropropene)," International Journal of Refrigeration, Vol. 33, No. 3, 2010, pp. 474-479.

13. Y. Higashi and K. Tanaka, "Critical Parameters and Saturated Densities in the Critical Region for trans-1,3,3,3-Tetrafluoropropene (HFO-1234ze(E))," Journal of Chemical & Engineering Data, Vol. 55, No. 4, 2010, pp. 1594-1597.

14. USEPA, "Transitioning to Low-GWP Alternatives in Unitary Air Conditioning," US Environmental Protection Agency, 2010. http://www.epa.gov/ozone/downloads/EPA_HFC_UAC.pdf

15. D. Del Col, D. Torresin and A. Cavallini, "Heat Transfer and Pressure Drop during Condensation of the Low GWP

Refrigerant R1234yf," International Journal of Refrigeration, Vol. 33, No. 7, 2010, pp. 1307-1318.

16. R. Akasaka, "An Application of the Extended Corresponding States Model to Thermodynamic Property Calculations for trans-1,3,3,3-tetrafluoropropene (HFO-1234 ze(E))," International Journal of Refrigeration, Vol. 33, No. 5, 2010, pp. 907-914.

17. R. Akasaka, K. Tanaka and Y. Higashi, "Thermodynamic Property Modeling for 2,3,3,3-tetrafluoropropene (HFO-1234yf)," International Journal of Refrigeration, Vol. 33, No. 1, 2010, pp. 52-60.

18. AHRI, "ANSI/AHRI Standard 870-2005 with Addendum 1, Performance Rating of Direct GeoExchange Heat Pumps," Air-Conditioning, Heating, and Refrigeration Institute, Arlington, 2009.

2

CHAPTER

THE DESIGN OF HEAT EXCHANGERS

Arturo Reyes-León[1], Miguel Toledo Velázquez[1], Pedro Quinto-Diez[1], Florencio Sánchez-Silva[1], Juan Abugaber-Francis[1], Celerino Reséndiz-Rosas[2]

[1]*Applied Thermal and Hydraulic Engineering Laboratory SEPI-ESIME-IPN Professional Unit, Lindavista, México D.F.*

[2]*División de Estudios de Posgrado e Investigación, Instituto Tecnológico de Pachuca, Pachuca de Soto Hidalgo, México*

ABSTRACT

A relation between heat transferred and energy loss, for turbulent flow. In different tube arrangements, is made. The conditions are determined which decide the dimensions and velocities for a heat exchanger. Also, a reference to the economic dimensioning of heat exchangers is presented. In this study, the conditions which a heat exchanger must satisfy represent the best balance between the amounts of material employed. The investigation is restricted to the case of turbulent flow.

Keywords: Heat Changers, Energy Loss, Economic Dimensioning, Counterflow, Crossflow and Parallelflow

1. INTRODUCTION

The heat exchanger is an equipment that allows heat transference between two fluids at different temperatures. Heat exchangers are extensively used in industry due to their wide variety of construction and applications in heat transference processes for producing conventional energy such as condensers, heaters, boilers or steam generators. They provide an adequate surface for heat transference to occur and their mechanical and thermal characteristics allow high pressure and high temperature processes.

Heat exchangers are important, their optimization rises the competitiveness and allows energy saving. The necessity of saving and recovering energy for different processes in industry makes essential the develop of new manufacturing technology for heat exchangers in order to cover a wide range of operation conditions. In recent years, new software for design heat exchangers has been focus in adapting the equipment to the required process and new solutions have been found that make the design time shorter.

2. HEAT TRANSFER AND FRICTION WORK

Let us first seek a relation between the heat transfer and energy loss. The problem is to determine the energy E which has to be spent in order that a quantity of heat Q may be transferred to a surface F, the average temperature difference being Q [1,2].

Symbols used:

Q = quantity of heat flowing per second.

G = weight of gas or liquid flowing per second.

Q = heat transfer coefficient.

Cp = specific heat at constant pressure.

Δt = temperature variation of medium during flow.

w = velocity of flow.

ζ = coefficient in pressure drop formula

l = tube length.

d = tube diameter.

γ = specific weight.

μ = absolute viscosity.

λ = thermal conductivity.

System units: m, kg, sec.

The heat given up to the surface F is:

$$Q = \alpha \cdot \Theta \cdot F \tag{1}$$

And the gas loses a corresponding amount of heat

$$Q = G \cdot C_p \cdot \Delta t \tag{2}$$

whence

$$\frac{\Delta t}{\Theta} = \frac{\alpha F}{G C_p} \tag{3}$$

It is convenient to derive the expression of the energy loss first for longitudinal flow through the tubes, and applies it to the case of cross flow over a tube bank. The expression for the pressure drop in a tube is

$$\Delta p = \zeta \frac{1}{d} \cdot \frac{w^2}{2g} \gamma \quad (4)$$

Putting the free gas section equal to f, and remembering that $G = w\gamma f$ and further that $\frac{1}{d} = \frac{F}{4f}$ we obtain from Equation (3)

$$\frac{\Delta p}{\Delta t} = \zeta \frac{w^3 \gamma^2 Cp}{2g\alpha} \quad (5)$$

The ratio of the pressure drop in the heat transfer depends, therefore, of the velocity with which the heating surfaces are swept. If we group together those quantities which depend on the velocity, and introduce the Reynolds number $Re = \frac{w\gamma d}{\mu g}$ and the Nusselt number $Nu = \frac{\alpha d}{\lambda}$ we obtain the following relation between heat transfer and pressure drop:

$$\frac{\Delta p}{\Delta t/\Theta} = \frac{\mu^3 g^2 Cp}{8\gamma\lambda d^2} \cdot \zeta \cdot \frac{Re^3}{Nu}.$$

utting also:

$$\zeta \frac{Re^3}{Nu} = Z, \quad (6)$$

enables us to write the desired relation between the heat transfer and the energy loss in the case of longi tudinal flow through a tube, referred to 1 kg of medium flowing through the exchanger as

$$\frac{\varepsilon}{\Delta t/\Theta} = \frac{\mu^3 g^2 Cp}{8\gamma\lambda d^2} \cdot z \quad (7)$$

Similar expressions may be obtained for the case of a tube bank with cross flow, if ζ is taken as denoting the pressure drop coefficient per tube row, if the bank is z_1 the expression for the pressure drop becomes

$$\Delta p = \zeta \cdot z_1 \cdot \frac{w^2}{2g}\gamma \tag{4a}$$

where w is the velocity at the narrowest point between the tubes. If $s \cdot d$ denotes the pitch of the tubes across the flow, and z_q denotes the number of tubes per row also in the direction across the flow, then with $F = \pi d l z_q \cdot z_1$ we obtain

$$\frac{\Delta p}{\Delta t/\Theta} = \frac{w^3\gamma^2 Cp}{8g\alpha} \cdot (s - \ ) \cdot \frac{4}{\pi}\zeta \tag{5a}$$

and introducing Re and Nu gives

$$\frac{\Delta p}{\Delta t/\Theta} = \frac{w^3 g^2 Cp}{8\gamma\lambda d^2} \cdot (s-1) \cdot \frac{4}{\pi} \cdot \zeta \cdot \frac{Re^3}{Nu}$$

Let us put

$$(s-1)\frac{4}{\pi} \cdot \zeta \cdot \frac{Re^3}{Nu} = Z_q \tag{6a}$$

and we obtain a new form of Equation (7), for the case of cross flow:

$$\frac{E}{\Delta t/\Theta} = \frac{\mu^3 g^3 Cp}{8\gamma^2 \lambda d^2} \cdot Z_q \tag{7a}$$

giving the relation between the energy loss and the heat transfer for cross flow.

It is seen that the first term of the Equations (7) and (7a) contains characteristic quantities of the medium and the tube diameter. This means that for a given tube diameter the heat exchanger is fully characterized by the number Z. Now the pressure drop coefficient ζ is function of Re, while Nu is a function of Re and of the Prandtl number, if we neglect the transition zone at the inlet, which is entirely permissible with cross flow exchangers many rows deep, or with longitudinal flow exchangers with relatively long tubes. But since the Prandtl number is purely a function of characteristic quantities of the medium, and we are interested only in a comparison of heat exchangers working with the same medium, and operating within the same temperature limits, Nu depends only on Re. This makes it possible to plot Nu as a function of Z. The Nu-Z diagram, therefore, gives a clear picture of the merit of a heat exchanger surface. Tube banks of different pitch are represented by different curves in the Nu-Z diagram. The higher a curve lies, the greater may be the heat transfer loading for a given energy expenditure, or, conversely, for a given surface and heat loading, the smaller the energy expenditure. Figure 1 which is drawn for gases contains curves relating to some of the most frequently used tube arrangements. In the case of cross flow, they are based on the values derived by Grimison from the tests of Pierson and Huge.

In order to make a comparison, curves for longitudinal flow have been inserted calculated with the aid of the formula given by Jung. It is seen that only at very high rates of heat transfer such as those which are achieved by the gas velocities attained in the Velox boiler, the longitudinal arrangement become more advantageous than the cross flow one. The curves giving the values of Z which are plotted in Figure 1, in a logarithmic scale is almost straight. It is, therefore, permissible when considering segments of these curves, and without introducing any appreciable error to assume the following relation

$$\log Z = B + m \log Nu$$

or,

$$Z = BNu^{m} \tag{8}$$

where B and m are constants whose value depends on the position in the diagram of the segment under consideration.

The expression for power loss can be written using Equation (7) and transforming Equation (3):

$$L = \frac{\mu^3 g^2}{8\gamma^2 d^3} \cdot F \cdot B \cdot Nu^{m+1} \tag{9}$$

1. Crossflow, staggered tubes, pitch 1.25 × 1.25.
2. Crossflow, tubes in line, pitch 1.25 × 1.25.
3. Crossflow, tubes in line, pitch 1.5 × 1.5.
4. Crossflow, tubes in line, pitch 2 × 2
5. Crosssflow, tubes in line, pitch 3 × 3.
6. Longitudinal flow in a tube.
7. Longitudinal flow between tubes, pitch 1.5 × 1 .5.
8. Longitudinal tinybetween tubes, pitch 2 × 2.
9. Single tube in crossflow.

The pitch is expressed as times the tubes diameter.

The diagram shows for some common tube arrangements, the relation between the heat transfer number No and characteristic number Z for the energy loss. For any given arrangement of the tubes, there is a definite value of Z for every value of Nu of heat transfer, with the help of which the pressure drop and the exchanger surface may be obtained from the Equation (7) or (7a) [3].

3. THE CONDITION FOR THE RIGHT EXCHANGER

The merit of an exchanger can only be judged when it is known what quantity of heat is equivalent to the mechanical energy which has to be supplied in the form of compressor or pump work to overcome the resistance of the exchanger. If the exchanger is an air preheater forming part of a steam power unit or a gas turbine then the overall efficiency of the plant or the efficiency when the exchanger is in service, determines the amount of the energy (expressed in kcal), to the heat consumption required to produce this energy [4,5].

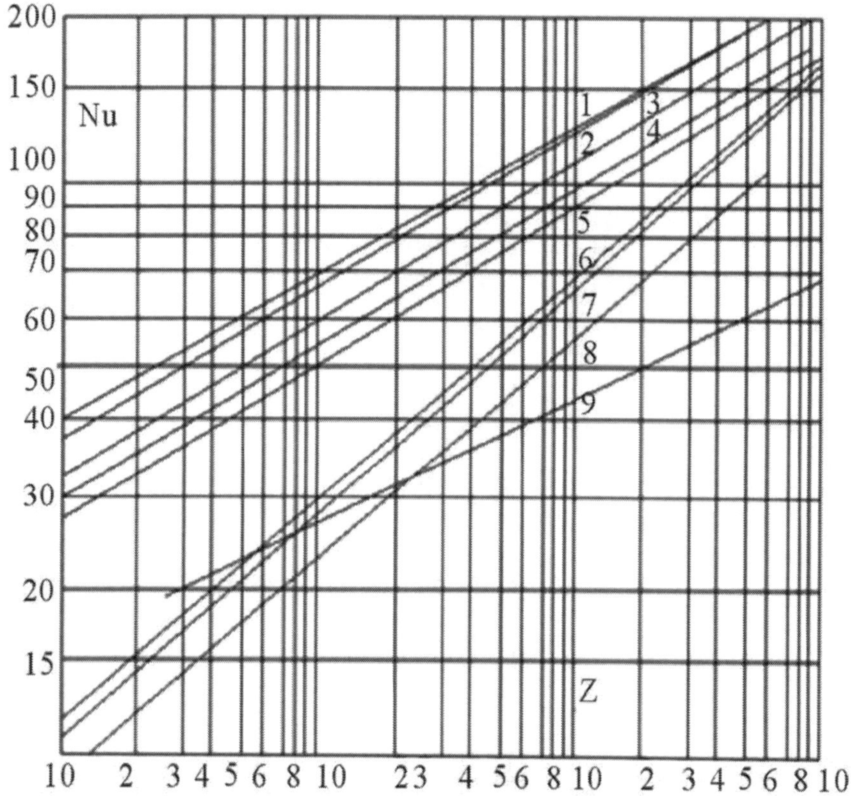

Figure 1. Relation between the energy loss and the heat transfer.

On the other hand, if it is a case of plain heat transfer for instance, in blast heaters, furnaces, etc., where the energy absorbed in overcoming the resistance of the exchanger has to be supplied in the form of power purchased from an outside supply, then the cost of this power must be balanced against the cost of production of the more or less completely transferred heat in the exchanger. If h denotes the efficiency of the plant with which the heat energy of the fuel is converted into mechanical energy, then the economic performance of the exchanger is given by

$$Useful = Q - \frac{AL}{\eta}.$$

useful heat is, therefore, equal to the difference of the heat transferred and the heat required for the production of the mechanical work absorbed.

The right heat exchanger is, therefore, the exchanger which with a given surface and with a given diameter of tubes results in a maximum amount of useful heat. The condition for this is

$$\eta dQ - AdL = 0 \qquad (10)$$

We shall now seek expressions for dQ and dL in terms of Nu.

Let the suffix 1 denote the hot medium, and the suffix 2 the cold one. The meaning of the symbols is made clear by the Figure 2. We may write for the temperature variation of the hot medium

$$t_1' - t_1'' = \varepsilon_1 \cdot (t_1' - t_2').$$

and similarly, that of the colder one is

$$t_2' - t_2'' = \varepsilon_2 \cdot (t_1' - t_2').$$

Further, let the mean temperature difference be given by

$$\Theta = a \cdot (t_1' - t_2').$$

The factor a depends only on e_1 and e_2, and is plotted in Figure 3 for counter flow, for cross flow and parallel flow, for the case $e_1 = e_2$. If we denote by k the overall heat transfer coefficient, the heat transferred is given by

$$Q = a \cdot k \cdot F(t_1' - t_2') \qquad (11)$$

For a small change in the rate of heat transfer we have.

$$dQ = \left(a + kF \frac{da}{dkF} \right) \cdot dkF \cdot \left(t_1' - t_2' \right)$$

Since kF may be treated as a single quantity.

We put

$$dQ = b \cdot dkF \cdot \left(t_1' - t_2' \right) \tag{12}$$

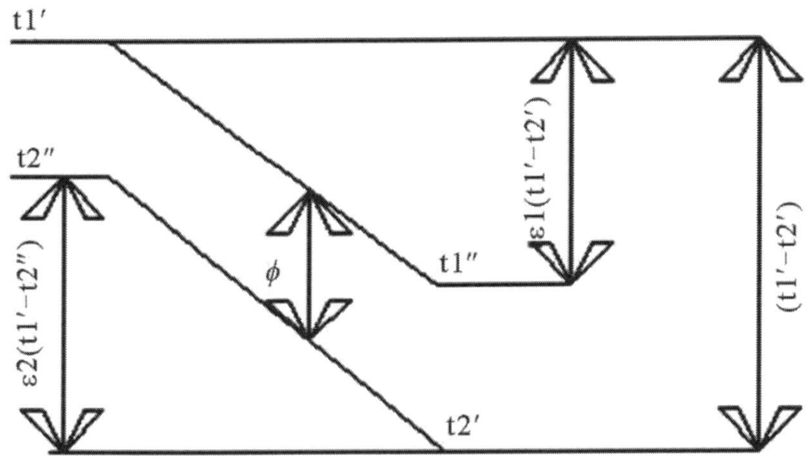

Figure 2. Diagram of temperature variation in an exchanger.

where b depends only on e_1 and e_2. From Equations (11) and (12), and from the relation $Q = \varepsilon GCp\left(t_1' - t_2'\right)$, it is found that

$$b = \frac{a}{1 - \dfrac{\varepsilon}{a}\dfrac{da}{d\varepsilon}} \tag{13}$$

Curves for b are given in Figure 3. In the case of heat transfer in metal exchangers the thermal resistance of the exchanger wall may be neglected without introducing any appreciable error. Hence, it is permissible to write

$$\frac{1}{kF} = \frac{1}{\alpha_1 F_1} + \frac{1}{\alpha_2 F_2} \tag{14}$$

Differentiating and introducing the Nusselt number in place of the quantities $d\alpha_1$ and $d\alpha_2$ gives

$$dkF = \left(\frac{kF}{\alpha_1 F_1}\right)^2 \cdot F_1 \cdot \frac{\lambda_1}{d_1} \cdot dNu_1 + \left(\frac{kF}{\alpha_2 F_2}\right)^2 \cdot F_2 \cdot \frac{\lambda_2}{d_2} \cdot dNu_2 \tag{15}$$

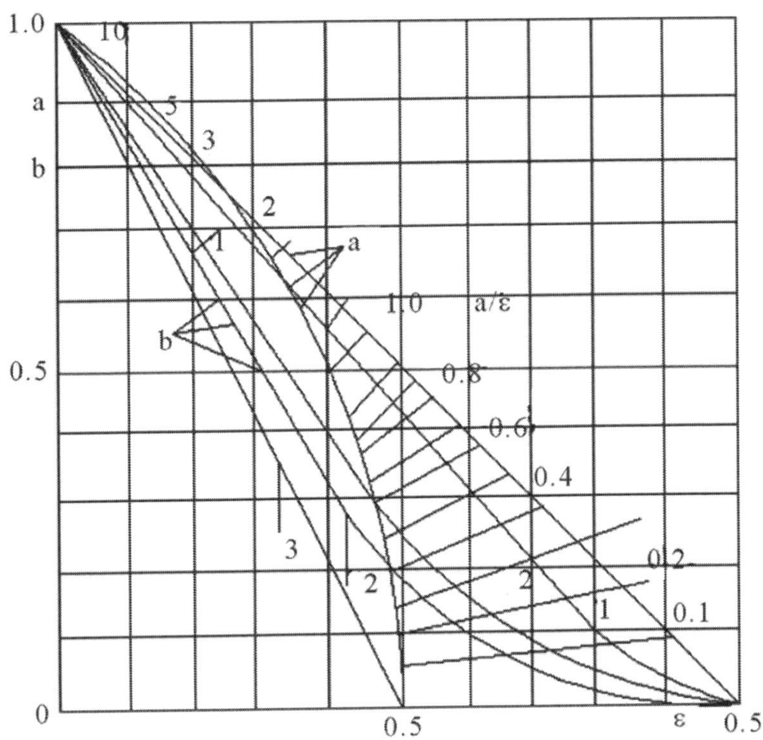

Figure 3. Characteristic numbers a and b of Equations (11) and (12) for $e_1 = e_2$ in the case of counterflow, crossflow and parallelflow.

The total power loss is the sum of the losses for the hot and cold mediums. We use Equation (9) and put

$$F\frac{\mu^3 g^2 B}{8\gamma^2 d^3} = P \tag{16}$$

and obtain for the total energy loss

$$L = P_1 Nu_1^{m_1+1} + P_2 Nu_2^{m_2+1}.$$

The change in loss with change of velocity is then given by

$$dL = P_1(m_1+1)_1\ Nu_1^{m_1} dNu_1 + P_2(m_2+1) Nu_2^{m_2} dNu_2 \tag{17}$$

The change in heat quantity transferred with change of specific loading kF is given by Equation (12). Inserting Equations (12) and (17) in Equation (10) and taking into account Equation (15) gives

$$\frac{\eta b}{A}(t_1' - t_2')\left[\left(\frac{kF}{\alpha_1 F_1}\right)^2 \cdot F_1 \cdot \frac{\lambda_1}{d_1} \cdot dNu_1 + \left(\frac{kF}{\alpha_2 F_2}\right)^2 \cdot F_2 \cdot \frac{\lambda_2}{d_2} \cdot dNu_3\right] - P_1(m_1+1) Nu_1^{m_1} dNu_1 - P_2(m_2+1) Nu_2^{m_2} dNu_2 = 0 \tag{18}$$

The coefficients of dNu_1 and dNu_2 must be equal 0. This gives two new equations namely,

$$\frac{\eta b}{A}(t_1' - t_2')\left[\left(\frac{kF}{\alpha_1 F_1}\right)^2 \cdot F_1 \cdot \frac{\lambda_1}{d_1}\right] = P_1(m_1+1) Nu_1^{m_1} \tag{19a}$$

$$\frac{\eta b}{A}(t_1'-t_2')\left[\left(\frac{kF}{\alpha_2 F_2}\right)^2 \cdot F_2 \cdot \frac{\lambda_2}{d_2}\right] = P_2(m_2+1)Nu_2^{m_2}$$

(19b)

Dividing Equation 19 (b) by Equation 19(a), inserting for P_1 and P_2 the values given by Equation (16) and substituting Nu for a_1 and a_2 gives

$$\frac{Nu_2^{m_2+2}}{Nu_1^{m_1+2}} = \frac{(m_1+1)\cdot B_1 \cdot F_1^2 \mu_1^3 \gamma_2^2 \lambda_1 d_2^4}{(m_2+1)\cdot B_2 \cdot F_2^2 \mu_2^3 \gamma_1^2 \lambda_2 d_1^4}$$

(20)

This equation determines the ratio of the velocities in the right heat exchanger; it does not, however, say anything about the absolute value of the velocities. It means that heat exchangers in which this ratio of the velocities is observed have for a given surface and a given heat quantity the lowest friction loss.

If we take the roots of Equations 19(a) and 19(b), and remembering that $\frac{kF}{\alpha_1 F_1} + \frac{kF}{\alpha_2 F_2} = 1$ we obtain as the second condition for the right heat exchanger:

$$\sqrt{\frac{\eta}{A} b(t_1'-t_2')} = \sqrt{R_1 Nu_1^{m_1}} + \sqrt{R_2 Nu_2^{m_2}}$$

(21)

The factor R contains only constants.

$$R = (m+1)\frac{\mu^3 g^2 B}{8\gamma^2 d^2 \lambda}.$$

Two Equations (20) and (21) completely determine the most favourable heat exchanger. The resulting transcenddental equation must be solved by trial. The surfaces are found with the aid of the numbers Nu, and Nu, and since Nu is a function of the Reynolds number, the velocities w_1 and w_2 are also determinate. The dimensions of the heat exchanger are, therefore, fixed.

4. THE ECONOMIC DIMENSIONING OF A HEAT EXCHANGER

It was seen in the first part that there is a function Z which serves as a criterion of the merit of tube arrangements in heat exchangers. In the second part the conditions fix the dimensions of the surface and sections of the right heat exchanger. The exchanger should, however, like every other apparatus be correctly dimensioned from the economic point of view, that is the total sum of the capital charges and of the running costs should be a minimum.

If P denotes the capital cost, n the interest and deprecition rate, then the capital chargers are

$$K_1 = nP \, .$$

and if L is the power absorbed, h the efficiency, h the operating hours in one year and p the price per kilowatt-hour, then the power costs are

$$K_2 = Lh\frac{1}{\eta}p \, .$$

and the total costs in one operational year

$$K_1 + K_2 = nP + h\frac{1}{\eta}pL \, .$$

there should be a minimum hence,

$$nP + h\frac{1}{\eta}pL = 0 \tag{22}$$

The capital costs will increase approximately in proportion to the exchanger surface, and for a given tube diameter, inversely proportionally to the heat transfer number, or

$$F = Nu^{-1}\ ,\ P = Nu^{-1}.$$

But according to Equation (9), the energy proportional to F Nu^{m+1}; hence substituting for $F = Nu^{-1}$

$$L = CNu^{m}\ \text{or}\ P \,\square\, L^{-\frac{1}{m}}.$$

differentiating and dividing by P

$$\frac{dP}{P} + \frac{1}{m}\frac{dL}{L} = 0 \tag{23}$$

Dividing Equation (22) by Equation (23)

$$nP + h\frac{1}{\eta}pmL = 0 \tag{24}$$

The total costs in an operational year are a minimum when the capital charges amount to m-times the power costs. Within the range of practical application, that is, for Nu = 40 to 120 the curve which is over this range can be seen upon as a straight line, for instance curve 3 (cross flow heat exchanger with a tube pitch 1.5 × 1.5) gives an exponent m = 3.84 and a constant B = 166; for curve 7 (longitudinal flow with a tube pitch 1.5 d) the figures are m = 2.67 and B = 99 500.

The starting point for this study has been the as sumption of a fixed tube diameter and tube pitch. These and the choice between staggered or straight arrangement of the tubes are determined by dirt deposit and cleaning considerations. How closely these assumptions and the results of the calculation of the right heat exchanger may be adhered to in practice depends on manufacturing conditions, but in any case the above exposition serves as a guide to show in what direction and to what extent modifications are desirable.

5. EXAMPLE

Designing a compact heat exchanger for heating wings continuous 20,000 kg/h of air from an inlet temperature (T_{iair}) 10°C to an out temperature (air $T_{o\ air}$) of 55°C heating fluid used water volumetric flow $\dot{V}_w$ (10 m^3/h at temperature T of 95°C, the input date are shown in Table 1 [6,7].

The density and heat capacity of both fluids are obtained from tables to standard atmosphere conditions. The characteristic of tube and of the wins are shown in Table 2.

Diagram inlets from primary and secondary of the compact heat exchanger are show in Figure 4.

Unit thermal water consumption

$$q_{t\ agua} = \dot{m}_{agua} cp_{agua} = \dot{V}_{agua} \rho_{agua} cp_{agua}$$

$$q_{t\ agua} = 10\frac{m^3}{h} \times \frac{1h}{3600s} \times 962.036\frac{kg}{m^3} \times 4217\frac{J}{kgC}$$

$$= 11269,18\frac{W}{0_C} = 11.269\frac{kW}{0_C}$$

Unit thermal air consumption

$$q_{t\ aire} = \dot{m}_{aire} cp_{aire}$$

$$q_{t\ aire} = 20000\frac{kg}{h} \times \frac{1h}{3600s} \times 1.005\frac{kJ}{kgK}$$

$$= 5.577\frac{kW}{C}$$

$$T_{m\ aire} = \frac{55+10}{2} = 32.5°C$$

Table 1. Input data of the compact heat exchanger.

Input data		
T_{inlet} of Secondary fluid (Tfi)	10.00	°C
T_{out} of Secondary fluid (Tfo)	55.00	°C
T_{inlet} of Primary fluid (Tci)	95.00	°C
velocity primary fluid (Wp)	1.00	m/s
velocity secondary fluid (Ws)	3.00	m/s
Mass flow of primary fluid (mp)	2.77	kg/s
Mass flow of secondary fluid (ms)	5.55	kg/s
heat capacity primary fluid (Cpp)	4217	J/kg-°C
heat capacity secondary fluid (Cps)	1005.00	J/kg-°C
density primary fluid (ρp)	962.04	kg/m^3
density secondary fluid (ρs)	1.15	kg/m^3

Table 2. Characteristic of tube and of the wing.

Characteristic of tube		
Outside diameter of tube (do)	0.012700	m
Inside diameter of tube (di)	0.011280	m
Wheelbase of the tubes in the direction of the tubes (l)	0.080	m
Wheelbase of the tubes in the direction of height (h)	0.040	m
Area inside of the tube (At-int)	9.993E - 505	m^2
Characteristics of the wing		
Thickness of the wing (e)	3.00E - 04	m
Thermal conductivity (Aluminum , k)	203.52	W/m°C

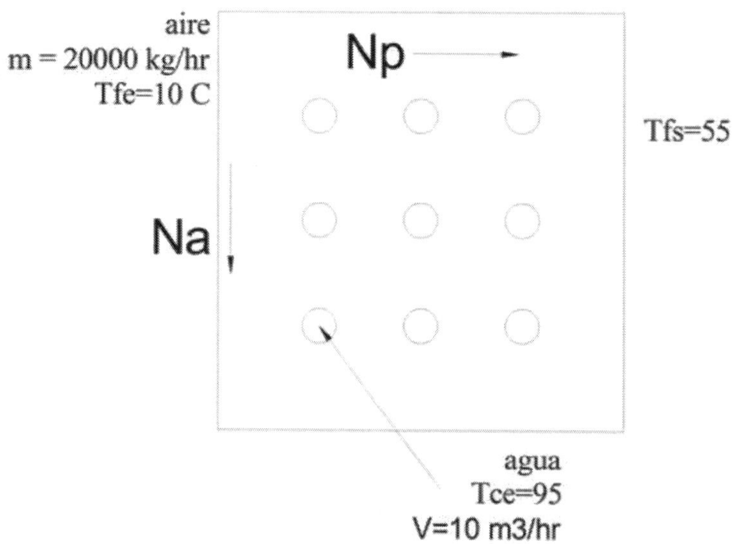

Figure 4. Diagram inlets from primary and secondary fluids.

The density of air to Tm is show in Table 3.

Table 3. Density of air to Tm.

T(°C)	$\rho\left(\frac{kg}{m^3}\right)$
30	1.165
32.5	$\rho_{air} = 1.15\left(\frac{kg}{m^3}\right)$
40	1.128

Themal flow

$$\dot{Q} = q_{\mathrm{t\,air}}\left(\mathrm{T}_{\mathrm{s\,air}} - \mathrm{T}_{\mathrm{e\,air}}\right) = \dot{m}_{\mathrm{air}} cp_{\mathrm{air}}\left(\mathrm{T}_{\mathrm{s\,air}} - \mathrm{T}_{\mathrm{e\,air}}\right)$$

$$\dot{Q} = 5577{,}75\frac{\mathrm{W}}{^{\circ}\mathrm{C}}$$

$$\left(55^{\circ}\mathrm{C} - 10^{\circ}\mathrm{C}\right) = 250.99KW = 250998{,}75W$$

Temperature of exit of the water

$$\mathrm{T}_{\mathrm{s\,water}} = \mathrm{T}_{\mathrm{e\,water}} - \frac{\dot{Q}}{q_{\mathrm{t\,water}}}$$

$$= 95^{\circ}\mathrm{C} - \frac{250998{,}75\mathrm{W}}{11269{,}18\,\mathrm{W}/^{\circ}\mathrm{C}} = 73.51^{\circ}\mathrm{C}$$

Velocities are: 2.5 m/s and 3.5 m/s

$$W_{\mathrm{water}} = 1\ \mathrm{m/s}$$

$$W_{\mathrm{air}} = 3\ \mathrm{m/s}$$

Logarithmic mean temperature

$$\Delta T_{ml} = \frac{\Delta\theta_1 - \Delta\theta_2}{\ln\frac{\Delta\theta_1}{\Delta\theta_2}} = \frac{(95-55)-(73.51-10)}{\ln\frac{40}{63.51}}$$

$$\Delta T_{ml} = 50.85^{\circ}\mathrm{C}$$

Surface airflow

$$S_p = \frac{\dot{V}_{\mathrm{air}}}{W_{\mathrm{air}}}$$

$$\dot{V}_{air} = \frac{W_{air}}{\rho_{air}}$$

Table gives the density of dry air at a temperature of 58.38°C

$$\rho_{air} = 1.15\frac{kg}{m^3}$$

Volumetric air flow

$$\dot{V}_{air} = \frac{20000\frac{kg}{h} \times \frac{1h}{3600s}}{1.15\frac{kg}{m^3}} = 4.83\frac{m^3}{s}$$

$$S_p = \frac{4.83\frac{m^3}{s}}{3\frac{m}{s}} = 1.61m^2$$

$$\frac{d_{int}}{d_{out}} = \frac{11.66mm}{15.88mm} \qquad h = 40mm$$

Surface water flow

$$S_T = \frac{\pi}{4}d_{int}^{\ 2}N_A; \qquad W_{water} = 1\ \text{m/s}$$

$$S_T = \frac{\dot{V}_{water}}{\text{W}_{water}} = \frac{10\frac{m^3}{h} \times \frac{1h}{3600s}}{1m/s} = 2.88 \times 10^{-3}\ \text{m}^2$$

Number of tubes in the direction of height (H) heat exchanger

$$N_A = \frac{S_T}{A_{\text{tube,inlet.}}} = \frac{4}{\pi}\frac{S_T}{d_{\text{inlet}}^{2}}$$

$$= \frac{4 \times 2.78 \times 10^{-3}\,\text{m}^2}{\pi(0.01128)\text{m}^2} = 29\text{tubes}$$

Height of heat exchanger

$$H = h \cdot N_A = 0.040\ m \cdot 29 = 1.15\ m$$

Winged tube length

$$S_p = H \cdot La \therefore La = \frac{S_p}{H} = \frac{1.61m^2}{1.15m} = 1.40m$$

In Table 4 are shown calculated values of Na, H and L to the compact heat exchanger.

Wing thickness, this value is assumed by the types of materials available in the industry

$$e = 0.3mm$$

Calculation of Reynolds number (Re), Nusselt (Nu) and the convective coefficient (hi) of primary fluid

Whereas, dh = di; and the Prandtl Number (Prp), dynamic viscosity (mp) and conductivity of the primary fluid are determined from tables to the average temperature of primary fluid (Tmp). These values are shown in Table 5.

$$Rep = \frac{Wp \cdot dh}{\mu p}$$

$$Nu = 0.023 \cdot Rep^{0.8} \cdot Prp^{0.3}$$

$$hi = \frac{Nu \cdot kp}{dh}$$

The values of Reynolds number, Nusselt number and convection coefficient inside of primary fluid are shown in Table 6.

Calculation of the mass velocity and maximum mass velocity of the secondary fluid.

$$s = \frac{h}{de}$$

$$Wms = Ws \cdot \rho s$$

$$Wms' = Wms\left(\frac{s}{s-1}\right)$$

The values of mass velocity, maximum mass velocity and relationship s are shown in Table 7.

Calculation of hydraulic diameter (dh¢), Reynolds (Re) and convection coefficient (I) as a function the secondary fluid pitch between wing (p) Tables obtained forthe Prandtl number (Pr), the dynamic viscosity (ms) and thermal conductivity (ks) of secondary fluid at film temperature (Tmp). These values are shown in Table 8.

Table 4. Values of Na, H and L to the compact heat exchanger.

Number of tubes in direction of height (*Na*)	29	Tubos
Exchanger height (*H*)	1.15	m
Overall longitude of the tube (*L*)	1.40	m

Table 5. Values of Prp, ν and kp.

Prandtl number of primary fluid (*Prp*)	2.11	
Viscosity of primary fluid (ν)	3.50E - 07	m^2/s
Conductivity of primary fluid (*kp*)	0.68	w/m °C

Table 6. Values of Rep, Nup and hi to the compact heat exchanger.

Reynolds number of primary fluid (*Rep*)	32228.57	
Nusselt number of fluid primary (*Nup*)	116.31	
Convection coefficient inside (*hi*)	6972.23	W/m^2-K

Table 7. Values of s, Wms and Wms of the secondary fluid.

Relationship between h and the outer diameter (*s*)	2.36	
Mass velocity of secondary fluid (*Wms*)	3.45	kg/s-m^2
Maximus velocity of secondary fluid (*Wms'*)	5.98	kg/s-m^2

Table 8. Values of Tp, Pr, rs and ks of secondary fluid.

Film Temperature (*Tp*)	58.33	°C
Prantl Number of secondary fluid (*Prs*)	0.6963	
Dynamic Viscosity of secondary fluid (ρs)	0.0781	kg/m-s
Conductividad del flujo secundario (*ks*)	0.0287	w/m °C

$$dh' = 2p\left(1 - \frac{\pi \cdot de^2}{4h \cdot l} - \frac{e}{p}\right)$$

$$\mathrm{Re} = \frac{Wms' \cdot dh'}{\mu s}$$

$$Nu = 0.26 \cdot Re^{0.6} \cdot Prp^{1/3}$$

$$he = \frac{Nu \cdot ks}{dh'}$$

Table 9. show the values of m, dh', Re, Nu and he of the secondary fluid.

Calculation of equivalent diameter (dea) and efficiencies of wings. (Table 10)

In Table 11 show characteristics of wing, Aluminum considering.

$$dea = \frac{2 \cdot h \cdot l}{h + l}$$

$$\mu = \frac{dea}{de}$$

$$\beta = \sqrt{\frac{2 \cdot he}{ks \cdot e}}$$

$\left[de \cdot \beta \cdot \frac{\mu - 1}{2}\right]$, Relationship for the efficiencies.

Calculation of the ratio number of wings with borders with the total number of wings.

Considering that N = Na x Np where Na is the number of tubes in direction of height and Np is the number of tubes in the direction of secondary flow.

Number of wings without borders

$$_{*}N0 = Na \cdot Np - 2(Na + Np) + 4$$

Number of wings with a border

$$_{*}N1 = 2(Na + Np) - 8$$

Number of wings with two borders

$$_{*}N2 = 4$$

*Note: This means that the equations are modified as a function on Np. (Table 12)

$$\psi 0 = \frac{N0}{N} \quad \psi 1 = \frac{N1}{N} \quad \psi 2 = \frac{N2}{N}$$

Calculation of correction coefficients for the wings. (Table 13)

$$C_0 = 1$$

$$C_1 = 1 + 0.5\left[\frac{h_e}{k_{al}}\left(\frac{l}{h} + \frac{h}{l}\right)\right]^{1/4}$$

$$C_2 = 1 + \left[\frac{h_e}{k_{al}}\left(\frac{l + h}{h}\right)\right]^{1/3}$$

Calculation of the overall efficiency of the equivalent circular wing. (Table 14)

$$\eta g = \eta f\left[\frac{\psi_0}{C_0} + \frac{\psi_1}{C_1} + \frac{\psi_2}{C_2}\right]$$

Table 9. Values of m, dh¢, Re, Nu and he of the secondary fluid.

Pitch (*m*)	Hydraulic Diameter (*dh'*)	Reynolds Number secondary (*Re*)	Nusselt Number secondary (*Nu*)	Convective Coefficient secondary (*he*)
0.002	3.08E - 03	718.4305	11.923624	110.9873
0.003	4.92E - 03	1147.548	15.792099	92.027767
0.004	6.77E - 03	1576.665	19.108255	81.04597
0.005	8.61E - 03	2005.782	22.077386	73.606079
0.006	1.04E - 02	2434.899	24.800817	68.113749
0.007	1.23E - 02	2864.017	27.337746	63.831783
0.008	1.41E - 02	3293.134	29.726447	60.364755

Table 10. Values of equivalent diameter and efficiencies of wings.

Pitch (*m*)	Convective Coefficient secondary (*he*)	Equivalence Factor wings (β)	$dea^{**} = (de^{*}\beta)^{*}(\mu-1/2)$	Efficiency obtained from the graphic ηf
0.002	110.9873	60.295115	0.823028315	0.9
0.003	92.027767	54.904119	0.749441229	0.92
0.004	81.04597	51.524193	0.703305238	0.93
0.005	73.606079	49.102356	0.670247159	0.95
0.006	68.113749	47.234887	0.644756207	0.96
0.007	63.831783	45.726081	0.624161001	0.97
0.008	60.364755	44.466936	0.606973681	0.98

Table 11. Characteristics of Aluminum.

Thermal conductivity of aluminum (*k*)	203.525	w/mC
Wing thickness (*e*)	3.00E - 04	m

Table 12. Relations with number of wings on the total number of wings.

Relations with number of wings on the total number of wings						
ψ	$(Np) = 1$	(Np) = 2	(Np) = 3	(Np) = 4	(Np) = 5	(Np) = 6
$\tilde{\psi}_0$	0	0	0.3101952	0.4652928	0.5583514	0.620390394
ψ_1	0	0.930586	0.6435285	0.5	0.4138829	0.35647147
ψ_2	0.930585591	0.069414	0.0462763	0.0347072	0.0277658	0.023138136

Table 13. Correction factors for each type of wing.

Correction factors for each type of wing				
Pitch (m)	h_e	c_0	c_1	c_2
0.002	110.987	1	1.510965	2.029347
0.003	92.0278	1	1.487588	1.967038
0.004	81.046	1	1.472341	1.926932
0.005	73.6061	1	1.461107	1.897653
0.006	68.1137	1	1.452253	1.874747
0.007	63.8318	1	1.444972	1.856018
0.008	60.3648	1	1.438802	1.840231

Calculation the overall surface heat Exchange where, $X = Np \cdot l$. (Table 15)

$$Se = (La \cdot X \cdot H)\left(\left(\frac{\pi \cdot de}{h \cdot l}\right) + \frac{2}{p} - \frac{\pi \cdot de^2}{2 \cdot p \cdot h \cdot l}\right)$$

Calculating the overall coefficient of heat transfer function and pitch Np

$$U = \frac{1}{\frac{Se}{Si}\left(\frac{1}{hi} + \frac{et}{kt}\right) + \frac{1}{he}\left[1 - \frac{Ss}{Se}(1 - \eta g)\right]}$$

where,

$$Si = \pi \cdot dik \cdot La \cdot Na \cdot Np$$

$$Ss = 2 \cdot \frac{La}{p}\left(X \cdot H - Np \cdot Na \cdot \left[\frac{\pi \cdot de^2}{4} \right] \right)$$

*kt = conductivity of tube, e_t = Thickness of the wall of tube.

**The value $\frac{et}{kt}$ is neglected as being small compared to $\frac{1}{hi}$. (Table 16)

Calculation Heat flow based on Np and pitch (p)

$$Q = U \cdot A \cdot (\Delta T).$$

Table 14. Overall efficiency (hg) as a function of Np and pitch (p).

Overall efficiency (ηg) as a function of Np and pitch (p).						
ηf	$(Np)=1$	$(Np)=2$	$(Np)=3$	$(Np)=4$	$(Np)=5$	$(Np)=6$
0.9	0.412708	0.585084	0.683014	0.731979	0.761358	0.780944
0.92	0.435243	0.607987	0.705014	0.753528	0.782636	0.802041
0.93	0.449131	0.621303	0.717299	0.765297	0.794095	0.813294
0.95	0.465868	0.639809	0.736269	0.784499	0.813437	0.832729
0.96	0.476524	0.650701	0.746883	0.794974	0.823829	0.843066
0.97	0.486346	0.660974	0.757071	0.80512	0.833949	0.853168
0.98	0.495576	0.670808	0.766957	0.815031	0.843876	0.863105

Table 15. Total surface heat exchange.

TOTAL SURFACE HEAT EXCHANGE						
P(m)	*(Np)* = 1	*(Np)* = 2	*(Np)* = 3	*(Np)* = 4	*(Np)* = 5	*(Np)* = 6
0,002	60,85782	121,7156	182,5734	243,4313	304,2891	304,2891
0,003	41,10674	82,21349	123,3202	164,427	205,5337	205,5337
0,004	31,23121	62,46242	93,69363	124,9248	156,156	156,156
0,005	25,30589	50,61178	75,91766	101,2236	126,5294	126,5294
0,006	21,35567	42,71135	64,06702	85,42269	106,7784	106,7784
0,007	18,53409	37,06818	55,60228	74,13637	92,67046	92,67046
0,008	16,41791	32,83581	49,25372	65,67162	82,08953	82,08953

Table 16. Overall rate of heat transfer (W/m^2 C).

OVERALL RATE OF HEAT TRANSFER (W/m^2 C)						
P(m)	*(Np)* = 1	*(Np)* = 2	*(Np)* = 3	*(Np)* = 4	*(Np)* = 5	*(Np)* = 6
0.002	100.1748	86.99647	80.94672	78.22676	76.68079	84.68974
0.003	109.8194	91.66171	83.87258	80.4542	78.53373	84.50856
0.004	110.6942	90.50486	82.55818	78.52662	76.50163	81.06726
0.005	107.1119	86.58511	65.14468	74.68027	72.68163	76.02262
0.006	102.8529	82.72923	62.38494	71.19159	69.25976	71.86373
0.007	98.2215	78.86459	72.69445	67.83032	65.9839	68.03117
0.008	93.64498	75.20035	69.27388	64.71031	62.95396	64.56843

where A = Se = Overall surface heat exchange DT = Tml = Difference logarithmic mean temperature. (Table 17)

Calculation $U \cdot Np \cdot St_1$ for comparison of the relationship ($Q/°C$) as a function of Np and the pitch between wings.

$$Q = U \cdot Np \cdot St_1 \cdot (\Delta\theta_{ml})$$

$$\therefore U \cdot Np \cdot St_1 = \frac{Q}{\Delta\theta_{ml}} = \frac{250998,75}{50.85} = 4935,724\ W/°C$$

Where:

St = Overall surface heat exchange with respect to pitch

Np = Pitch number of tubes in the direction of seconddary flow. (Table 18)

Selection based on the results obtained in the above table $U \cdot Np \cdot St_1 = \frac{Q}{\Delta\theta_{ml}}$ can be compared to select the exchange with their respective sizes, data are compared

$$4935,724059 \approx 5300,392$$

This variation is important for any factor that our team needs to transfer more heat.

The selection of compact heat exchanger required to have 6 columns of tubes (Np = 6) in the direction of flow and pitch inlet wings of 8 mm, so X will be worth

$$X = 1xNp = (0.008)(6) = 0.18\text{m}$$

The dimensions of the compact heat exchanger will (L) x(H) x(X), (1.40 × 1.15 × 0.18) m.

Table 17. Values of heat flow (Q).

Heat flow (Q)						
$P(m)$	$(Np)=1$	$(Np)=2$	$(Np)=3$	$(Np)=4$	$(Np)=5$	$(Np)=6$
0.002	310024.2	538478.9	751549.4	968394.7	1186571	1310502
0.003	229568.8	383223.1	525987	672732.6	820842.8	883292.4
0.004	175806.2	287482.5	393360.6	498868.8	607505.5	643761.5
0.005	137841.4	222851.5	251502.6	384421.9	467667.2	489164.7
0.006	111699.4	179689.6	203252.1	309259.2	376084.1	390223.8
0.007	92576.03	148663.4	205548.6	255726.6	310956.7	320604.8
0.008	78184.91	125570.7	173511.9	216108.5	262803.7	269543.4

Table 18. Values of $U \cdot Np \cdot St_1$ (W/°C).

$U \cdot Np \cdot St_1$ (W/°C)						
$P(m)$	$(Np)=1$	$(Np)=2$	$(Np)=3$	$(Np)=4$	$(Np)=5$	$(Np)=6$
0,002	6096,42	10588,83	14778,72	19042,84	23333,13	25770,16
0,003	4514,317	7535,829	10343,19	13228,84	16141,33	17369,36
0,004	3457,113	5653,152	7735,176	9809,925	11946,19	12659,14
0,005	2710,561	4382,226	4945,632	7559,402	9196,366	9619,099
0,006	2196,494	3533,477	3996,817	6081,377	7395,444	7673,492
0,007	1820,446	2923,367	4041,977	5028,693	6114,758	6304,48
0,008	1537,454	2469,264	3411,996	4249,631	5167,861	5300,392

Table 19. Characteristics of the exchange.

CHARACTERISTICS OF THE EXCHANGE		
de:	Outside diameter of the tubes	12.7 mm
di:	Internal diameter of the tubes.	11.28 mm
La:	Wings tube length	1.40 m
L:	Overall length of the tube.	1.40 m
N_a:	Number of vertical tubes in the exchanger.	29 tubos
N_p:	Number of tubes in the direction of secondary flow.	6 tubos
N:	Overall number of tubes ($N = Na \cdot Np$).	174 tubos
h:	Wheelbase of the tubes in the vertical direction.	0.04 m
l:	Wheelbase of the tubes in the direction of secondary flow.	0.08m
n:	Number of wings $(n = La/p)$.	175 wings
p:	Wings pitch (from center to center of wings)	0.008 m
e:	Wing thickness.	0.3 mm
H:	Height of heat exchanger (continuous wing)	1.15 m
X:	Heat exchanger width (within the meaning of secondary flow).	0.18 m

With piping HWG 22 de = 12.7 mm y di = 11.28 mm.

The thick aluminum wings e = 0.3 mm and K = 203.52 W/m C. In Table 19 are shown characteristics of the exchanger.

6. CONCLUSIONS

This paper identifies the advantages of having the appropriate exchanger with working conditions, environmental conditions and economic aspects, it is also necessary to mention the following regarding the general utility of this work.

- In addition to the thermal design, mechanical design of heat exchangers is also a part of it. The mechanical design is done under the ASME Section VIII, which is entitled "Pressure Vessel Design"
- Although the subject of this work is the design of heat exchangers, which, as noted was achieved successfully, the utility of it is wider and there are several methods for the design of heat exchangers.
- While this paper addresses just one example of a heat exchanger, this could vary, the same results in the thermal design, these variations are affected primarily for economic reasons or space.

REFERENCES

1. D. Q. Kern, "Procesos de Transferencia de Calor," 31th Reimpresión, CECSA, México, 1999.

2. F. P. Incropera and D. P. Dewitt, "Fundamentals of Heat and Mass Transfer," 5 th Edition, Wiley, New York, 2002.

3. M. Serna and A. Jimenez, "A Compact Formulation of Bell-Delaware Method for Heat Exchanger Desing and Optimation," Chemical Engineering Research and Desing, Vol. 83, (A5), 2005, pp. 539-550.

4. R. K. Shah, P. Dusan and D. P. Sekulic, "Fundamentals of Heat Exchanger Design," John Wiley & Sons, 2003.

5. Z. H. Ayub, "A New Chart Method for Evaluating Single-Phase Shell Side Heat Transfer Coefficient in a Single Segmental Shell and Tube Heat Exchanger," Applied Thermal Engineering, Vol. 25, 2005, pp. 2412-2420.

6. Y. A. Kara and Ö. Güraras, "A Computer Program for shelland-tube Heat Exchanger," Applied Thermal Engineering, Vol. 24, 2004, pp 1797-1805.

7. R. Castillo, "Diseño Computacional de Intercambiadores de Calor de Coraza y Tubos por el Método Delaware," Tesis de Maestría, Instituto Politécnico Nacional, México, 1999.

3

CHAPTER

ZPPR FUEL ELEMENT THERMAL STRESS-STRAIN ANALYSIS

Charles W. Solbrig, Jason Andrus, Chad Pope

Idaho National Laboratory, Idaho Falls, USA

ABSTRACT

The design temperature of high plutonium concentration ZPPR fuel plates is 600°C. Cladding integrity of the 304 L stainless steel cladding is a significant concern with this fuel since even small holes can lead to substantial fuel degradation. Since the fuel has a higher coefficient of thermal expansion than the cladding, an investigation of the stress induced in the cladding due to the differential thermal expansion of fuel and cladding up to the design temperature was conducted. Small holes in the cladding envelope would be expected to lead to the fuel hydriding and oxidizing into a powder over a long period of time. This is the same type of chemical reaction chain that exists in the degradation of the high uranium concentration ZPPR fuel. Unfortunately, the uranium fuel was designed with vents which allowed this degradation to occur. The Pu cladding is sealed so only fuel with damaged cladding would be subject to this damage. The thermal stresses that can be developed in the fuel cladding have been calculated in this paper and compared to the ultimate tensile stress of the cladding. The conclusion is drawn that thermal stresses cannot induce holes in the cladding even for the highest storage temperatures predicted in calculations (292°C). In fact, thermal stress cannot cause cladding failure as long as the fuel temperatures are below the design limit of 600°C (1112°F).

Keywords: ZPPR Nuclear Fuel, Fuel Failure, Fuel Hydriding and Oxidizing, Thermal Induced Stresses and Strains, Plastic Deformation

1. INTRODUCTION

The purpose of this paper is to describe the stresses and strains and possible damage which can occur in the ZPPR fuel elements when raised to a given temperature to determine if damage to the cladding can occur. Damage to the cladding is a concern because even a small hole can allow air and moisture to be exposed to the fuel. The fuel will oscillate through the hydride oxide cycle (like the ZPPR Uranium fuel described in Ref. [1]) over a long period of time and eventually turn into an oxide powder. This turns the metal fuel into a useless highly reactive hydride powder that can burn and spread (contamination) quickly during handling. It would need to be processed back to a useable form. The conductivity of both the metal and the clad are so high that the entire fuel element would be at a uniform temperature in storage. The temperature which the fuel elements achieve in storage under various conditions is estimated in Ref. [2] . This paper concludes that no damage would occur to the fuel elements is raised to 290°C (554°F) (which is the maximum temperature calculated in Ref. [2]), because the maximum plastic strain that can be induced in clad due to the fact that this temperature is small, and even though it is possibly beyond the yield strain, it cannot cause a hole in the clad because the strain will not exceed 0.005 even though the elastic limit strain is 0.0012. Failure does not occur inface centered cubic (fcc) metals with elongation less than 0.20. No damage is predicted as long as the fuel temperature is below the design limit of 600°C. Thus, even though damage has occurred to a limited number of fuel elements [3] , this study concludes that this damage has not been caused by the temperatures encountered in storing the fuel.

A conservative calculation indicates the cladding would exceed its yield stress at any fuel element temperature above 170°C (338°F) (referred to as the yield temperature). A more realistic calculation indicates the fuel cladding yield temperature is reached at 370°C (698°F). At temperatures above the yield temperature, the clad will yield (i.e., permanently deform), but only a small amount because thermal expansion of the metal fuel is relatively small. The deformation will be small enough that the thinning would be in the behavioral region where the clad thickness thins uniformly with elongation and should not form any pinholes. Metal damage in fcc metal does not occur as long as the plastic strain less than 0.20 although most designs limit strains to significantly less than the elastic limit [4] .

This paper briefly describes the ZPPR reactor, then the fuel elements and the coefficient of thermal expansion of the fuel plate and the elastic plastic tensile characteristics of the clad. An elastic model of the fuel and cladding is developed, then a modification of the model is derived to describe the elastic-plastic behavior of the clad. The calculated results of the stress and strain using these models are compared to experimentally determined ultimate stresses and strains of the cladding.

1.1. Description of the ZPPR Reactor and Pu Based Fuel Elements

ZPPR (Zero Power Physics Reactor) was the largest of the split-table fast reactor critical facilities. ZPPR was operated as a criticality facility from April 18, 1969 until decommissioned in 1990. It was used to obtain a large amount of very detailed data on a variety of full-sized reactor configurations including large, commercial-sized fast reactors with powers of up to 1200 MW (electric). The purpose was to construct assemblies that closely resembled various fast reactor designs and then use the experimental results to validate and refine the data and methods used to design large fast reactors. There were 21 major assemblies, most of which had several major variants (e.g., beginning of cycle, middle of cycle, end of cycle, various control rod positions, etc.). The experiment campaigns included several large plutonium reactors, engineering mockups for both major design concepts for the Clinch River Breeder Reactor designs, larger fast reactor designs as part of the JUPITER collaboration with Japan, the SP-100 space reactor, and assemblies supporting the Integral Fast Reactor Design [5] .

The ZPPR critical assembly is put together from rectangular building blocks of fuel, coolant and structural materials inserted into drawers. Figure 1 shows a picture of the ZPPR reactor with a technician inserting material into the drawers in the left half of the reactor. The right half, also seen in Figure 1, contains identical drawers. The entire reactor system is shown in this figure. Once the two halves were fully assembled, all personnel would exit the room and the two halves would be slowly brought together using remote controls. With the reactor fully assembled, control rods would be carefully withdrawn and the reactor would be brought to a critical state. To avoid generating large quantities of fission products in the fuel plates and to prevent the need for coolant flow, a low reactor power would be maintained, typically on the order of one kilowatt. With the reactor in a low power critical state, various reactor physics parameters would be measured such as neutron reaction rates, neutron flux profiles, excess reactivity, and control rod worth measurements. The reactor is designed to withstand credible and incredible accidents and is surrounded by a containment designed with filters which would remove any radioactive contamination generated in a potential accident.

There is a very large inventory of critical facility materials that could be assembled in various combinations to construct any reactor within a short time. Because ZPPR was operated at very low power, the materials did not become very radioactive, and they were used over and over again and are inserted in the drawers by hand with minimal radiation protection required. These features, combined with the short time required to assemble a core, meant that nuclear reactors could be built and tested in ZPPR for about 0.1% of the capital cost of construction the whole power plant.

Figure 1. Experimenter loading fuel in the ZPPR reactor.

The cores were made up on many platelets of various materials which were loaded into drawers. The drawers were then loaded into a matrix structure which was made up of many square matrix tubes clamped tightly together into a large square array. The array of matrix tubes was positioned horizontally so that the drawers were inserted horizontally into the matrix tubes, and the core simulations were generally cylindrical with axis of the cylinder positioned horizontally. In addition, the array of matrix tubes was separated into two halves at the cylindrical axial midplane. During operation, the two halves of the array of matrix tubes were positioned tightly together, but at shut down the two halves of the assembly were separated giving easy access to the midplane of the assembly and provided a huge shut down reactivity margin.

The ZPPR Facility includes the ZPPR cell, ZPPR Vault/Workroom, the ZPPR Materials Control Building, and the ZPPR Control Room.

The fuel which is inserted in the drawers comes in thin 2 inch by 0.20 inch thick clad plates of different lengths up to 8 inches long. It was manufactured in 1967 and has been in storage since the reactor was decomissioned. So although the reactor has been decommissioned, the fuel must still be stored and maintained to prevent any degradation and possible contamination.

The high Pu ZPPR fuel plates are made of a Pu U Mo metalalloy [6] about 2 in. high, 0.2 in. thick, and of various lengths, (designated by A in Figure 2) from 1 to 8 in (See Figure 2). The plates are clad in a 304 L Stainless Steelsheet [7] wrapped around the plates with end caps which are inset into the clad wrapper and welded in. One end cap is welded into the cladding, then the fuel plate is inserted, and finally, the final seal is made by pressing the other end cap firmly against the fuel and then welding that end cap in place. The details of the fuel element geometry and assembly are described in more detail in the fuel specification [8] (PF-1600, 1967).

In the low temperature range, the coefficient of thermal expansion (CTE) of the fuel plate is the same as that of the cladding. At higher temperatures, the CTE of the fuel plate is greater than that of the cladding, so as the fuel element heats up, it will attempt to elongate more than the cladding will allow. Since the closure design of the clad requires both to be the same length at room temperature when the temperature rises, a tensile stress is built up in the cladding and a compressive stress in the fuel plate. The stresses and strains which build up as the fuel is heated up are modeled in this paper to determine if failure of the cladding due to thermal expansion is probable.

1.2. Coefficient of Thermal Expansion of the Fuel Plate

The coefficient of thermal expansion of the stainless steel is reported on the internet in several different references and is approximately $16 \times 10^{-6}/°C$ to $18.7 \times 10^{-6}/°C$ in the range of 0°C to 650°C with higher values being at the higher temperatures.

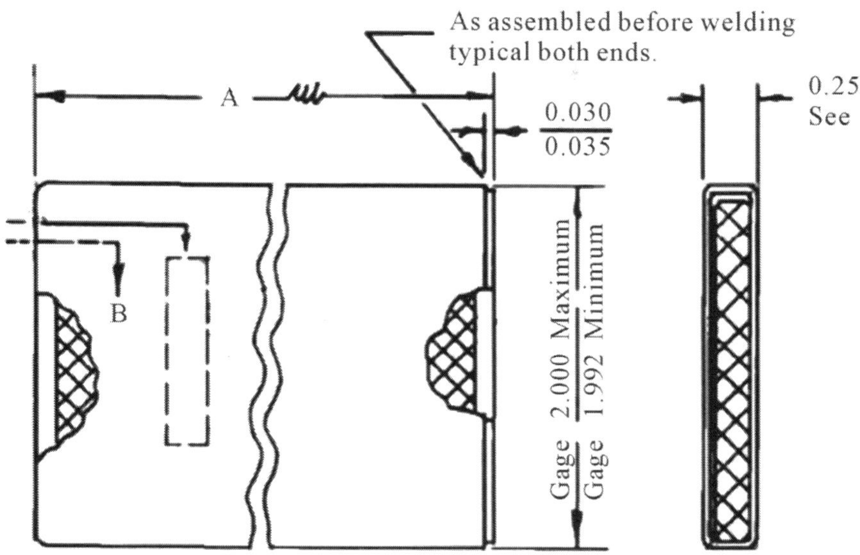

Figure 2. Schematic diagram of the ZPPR fuel elements.

The plutonium uranium alloy is more complicated. Plutonium undergoes more phase transitions at ordinary pressures than any other element. As plutonium is heated, it transforms through six different crystal structures, α, β, γ, Δ, Δ′, and ε, before melting at 640°C (1184°F). Density and thermal expansion vary significantly from phase to phase. The linear expansion of plutonium over all these phases [9] is shown in Figure 3. This figure shows that there are three volume expansion jumps in going through the first four phases. The first three phases show respectively CTEs of 48, 47, and 45×10^{-6} cm/cm°C. The fourth and fifth phases, Δ and Δ′, show negative CTEs and negative volume jumps between these two and the sixth phase. The sixth again shows an increase in CTE. The CTE and phase change volume jumps calculated from this graph are shown in Table 1.

When plutonium is mixed with uranium, the CTE changes quite a bit. Measurements made on the SEFOR fuel are shown in Table 2 which is taken from Johnson (1965) page 3-6. Dilatometric studies were made at Argonne National Laboratory in which 1 inch long cast specimens were heated and cooled through several cycles between 30°C (86°F) and 700°C (1292°F) to check for reproducibility. Average expansion coefficients for ranges of temperatures on heating are given in Table 2. The average coefficient increases from 15.3×10^{-6} cm/cm°C in the range 30°C (86°F) to 220°C (428°F) to 24.8×10^{-6} cm/cm°C in the range 380°C (716°F) to 545°C (1013°F). A transformation occurs between 545 (1013°F) and 617°C (1143°F) results in an expansion of 3.95×10^{-3} cm/cm but the average expansion rate is included in Table 2. As will be seen, these CTEs are similar to the that of the ZPPR fuel. According to Johnson [6] , the over-all expansion coefficient from 30°C (86°F) to 700°C (1292°F) is 24.6×10^{-6} cm/cm°C.

The compressive strength at room temperature and at 280°C (536°F) was 190,000 psi and 150,000 psi, respectively. It has enough toughness at room temperature to permit handling and machining but exhibited brittle failure in the above compression tests. Ductility increases rapidly above 400°C (752°F). Thin castings of the metal have smooth, bright surfaces. Porosity is not a problem. Castings can be drilled and machined with carbide tools.

The ZPPR FSAR [10] reports the CTEs for "Alloy Fuel" which are listed in Table 3. These values are higher than for the SEFOR fuel. Although not specified, it is assumed that the alloy fuel referenced in the FSAR is the NUMH fuel listed in Table 65 of Ref. [11] , which has a higher concentration of Pu (U-34 wt% Pu-2.5 wt% Mo-0.7 wt% Am) perhaps accounting for a higher CTE.

2. FUEL ELEMENT INTERNAL PRESSURE AND CLEARANCE BETWEEN FUEL AND CLAD

This section show that the fuel elements internal pressure is atmospheric even though the initial pressure loading is from 1/2 to 1/3 atmospheric pressure. It also shows that in most cases the cladding and the fuel are probably not rubbing against one another due to differential thermal expansion of the fuel and cladding.

The fuel element's final assembly consists of taking cladding which has one end plug already welded in it.

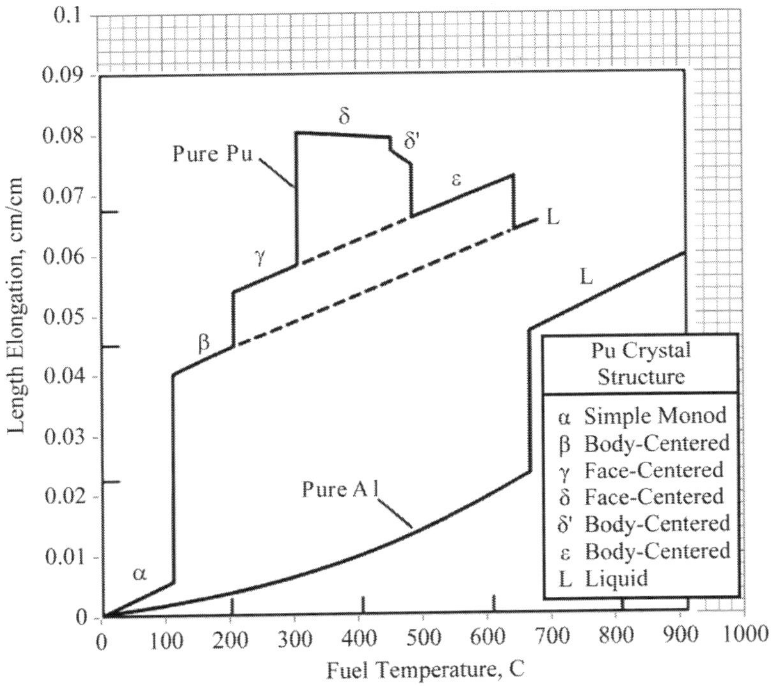

Figure 3. Length elongation for plutonium [9] .

Table 1.CTEs and phase change volume jumps from Figure 3.

Phase	Temperature °C	Length elongation %	CTE or phase change jump	Units
alpha	0 to 105	0.005	0.000048	cm/cm°C
alpha to beta	105	0.0355	0.0305	cm/cm
beta	105 to 200	0.04	0.000047	cm/cm°C
beta to gamma	200	0.0475	0.0075	cm/cm
gamma	200 to 300	0.052	0.000045	cm/cm°C
gamma to delta	300	0.0715	0.0195	cm/cm

Table 2. CTE for U-20 wt% Pu-2.5 wt% Mo SEFOR alloy casting.

Temp C	CTE	Units	
30 - 220	0.000015	cm/cm°C	
220 - 380	0.000024	cm/cm°C	
380 - 545	0.000025	cm/cm°C	
545 - 617	0.000055	cm/cm°C	Transformation
617 - 700	0.000023	cm/cm°C	

Table 3. CTE for alloy fuel [10] .

Temp °C	CTE	Units
27 - 227	0.000015	cm/cm°C
227 - 377	0.000029	cm/cm°C
377 - 545	0.000029	cm/cm°C
545 - 627	0.000052	cm/cm°C
627 - 957	0.000025	cm/cm°C

Then the fuel plate is inserted into the cladding and the other end plate is pressed into the cladding and held with a force of at least 40 lbf against the fuel while the end plug is welded in. The welding is done in a helium-argon atmosphere with at least 20% helium in a subatmospheric chamber between 1/3 and 1/2 atmosphere.

The specifications for the fuel element are summarized in this paragraph: "The welds at the ends of the plug and one side weld may be made with the welding chamber filled with helium-argon at atmospheric pressure. Before the final weld is made, the welding chamber shall be evacuated to a pressure of less than 0.2 torrfor a period of five minutes or more. A helium-argon welding gas mixture containing not more than 10 ppm H_2O and 30 ppm O_2 and not less than 20% helium is then admitted to the chamber to a pressure of 1/3 to 1/2 of a standard atmosphere (5 psia to 7.5 psia), and the final weld is made at this pressure."

After the fuel element closure was made and the fuel element was removed from the welding chamber and exposed to the atmosphere, the increase of pressure on the outside of the cladding to one atmosphere causes the cladding to collapse somewhat. As is shown in this section, this collapse will cause the pressure on the inside of the cladding to increase to atmospheric pressure. The volume of gas remaining in the cladding will be greater than the estimated minimum collapsed. The free volume between the fuel and cladding is referred to here as the plenum volume.

The calculations which demonstrate this for an 8-in.-long fuel element are described in this section. There is no pressure difference between the outside and inside of the cladding at the time the final welding is done, so the plenum volume at this time can be estimated by subtracting the volume of the fuel from the volume of the un-deformed cladding. Three different estimates of this uncollapsed plenum volume, the average, the maximum, and the minimum, are included in Table 4. The average value is obtained by using the average specification dimensions. The maximum is estimated by using the difference between the maximum cladding height specs and the minimum fuel height specs. The minimum is estimated by the difference between the minimum cladding height and the maximum fuel height. These dimensions and the volumes calculated from them are also included in Table 4. Note the cladding dimensions are inside dimensions, and all dimensions are at room temperature.

An estimate of the minimum volume remaining if the cladding were to completely collapse down onto the fuel, but not wrinkle, has been made and is also included in Table 4. This volume is termed the collapsed volume and is estimated by assuming that no volume exists between the cladding and the fuel at the end caps or between the cladding and the sides of the element which are about 8 in. long by 2 in. high but that the volume between the top of the fuel and cladding and between the bottom of the fuel element and cladding remain. With the 40 lbf applied at closure, the ends plugs press against the fuel and allow no space. The sides are pressed against the fuel by the outside pressure difference and allow no space on the sides. The top and bottom do not compress down on the fuel due to the stiffness of the metal cladding. This geometry is attested to from pictures of the fuel. If the top and bottom were pressed against the fuel, then wrinkles would have dented the cladding as evidence but none exist.

The collapsed volume that remains inside the cladding with the above model is estimated by the free volume on the top and bottom of the fuel which is estimated by the difference in the cladding height and the fuel height times the fuel thickness times the fuel length. This is included in Table 4 as the collapsed volume.

Table 4. The uncollapsed and collapsed fuel plenum volumes.

Parameter	Units	Initial Plenum Dimensions		
		Minimum	Average	Maximum
Fuel thickness	in	0.2000	0.2025	0.2050
Fuel hieght	in	1.9000	1.9300	1.9600
Clad thickness	in	0.2110	0.2120	0.2130
Clad height	in	1.9640	1.9650	1.9660
Volume inside clad	(in^3)	3.2332	3.2502	3.2671
Volume of fuel	(in^3)	3.1348	3.0492	2.9648
Uncollapsed Volume	(in^3)	0.0983	0.2009	0.3024
Collapsed Volume	(in^3)	0.0066	0.0576	0.1087

Using the perfect gas law, the volume remaining when the completed fuel elements are brought into the atmosphere at a pressure (13.4 psia) which is conservatively above the average local ZPPR pressure (12.3 psia) is shown in Table 5 for room temperature as well as the maximum temperature which has been determined that the ZPPR fuel could attain in storage. In all cases, these calculated volumes are greater than the minimum collapsed volumes. Thus, the cladding does not completely collapse onto the fuel, and the pressure inside the fuel elements is the same as the local atmospheric pressure. Note that Table 5 also shows the cladding does not completely collapse even at sea level (14.7 psia) but it almost collapses as is shown in Table 6. Note, the collapsed volumes in Table 4 are repeated in Table 5 for ease of comparison.

Table 6 shows the distance between the cladding and the fuel for different conditions. It is calculated by taking the volume in Table 5, subtracting the collapsed volume from it, dividing the remaining volume by the area of the fuel (15.3 in^2) and then dividing by 2 to assume a space on each side of the fuel. The stainless steel roughness is listed in The Engineering Toolbox [12] as 0.60 mils. If the fuel and cladding both have a roughness of 6 mils, then Table 6 shows those loaded at 5 psia are touching at room temperature. Under most conditions listed in Table 6, the friction would not be significant. The differential expansion of low temperature fuel will be very small so friction at these temperatures will also be insignificant.

3. METAL TENSILE STRESS-STRAIN FAILURE BEHAVIOR

The failure of the 304 L stainless steel cladding subjected to tensile stress when the fuel element temperature increases determines if the fuel is exposed to air and humidity. Its stress-strain-failure properties are presented in this section. Many metals subject to tensile stress follow a linear relationship between stress and stain (Hooke's Law) in the low stress strain region. This is termed the elastic region. After the limit of this region is reached, the material will distort in a non-linear manner by elongating more for a given increase in stress. The point at

Table 5. Fuel plenum volume (in^3).

Ambient Pressure (psia)	Fill Pressure (psia)	Fuel Temperature Temp °K	Fuel Temperature Temp °C	Initial Plenum Dimensions		
				Minimum	Average	Maximum
13.4	7.5	300	27	0.0550	0.1125	0.1692
13.4	5	300	27	0.0367	0.0750	0.1128
13.4	7.5	565	292	0.1037	0.2118	0.3188
13.4	5	565	292	0.0691	0.1412	0.2125
14.7	7.5	300	27	0.0502	0.1025	0.1543
14.7	5	300	27	0.0334	0.0683	0.1029
14.7	7.5	565	292	0.0945	0.1931	0.2906
14.7	5	565	292	0.0630	0.1287	0.1937
		collapsed	volume	0.0066	0.0579	0.1097

Table 6. Distance between cladding and fuel plate (mils).

Ambient Pressure (psia)	Fill Pressure (psia)	Fuel Temperature Temp °K	Fuel Temperature Temp °C	Initial Plenum Dimensions		
				Minimum	Average	Maximum
13.4	7.5	300	27	1.58	1.78	1.95
13.4	5	300	27	0.98	0.56	0.10
13.4	7.5	423	292	2.32	3.29	4.22
13.4	5	423	292	1.48	1.56	1.62
13.4	7.5	565	292	3.17	5.03	6.84
13.4	5	565	292	2.04	2.72	3.36

which straight line behavior ceases is called the limit of proportionality or elastic limit. Beyond this the material will not return to its original shape when the stress is released. This is called plastic behavior. Figure 4 as well as the next two paragraphs is from Hecker [4] . The elastic limit, the stress at which the material starts to exhibit permanent deformation, is also called the yield strength.

As Figure 4 shows, if the stress is increased beyond the yield point all the way to the ultimate tensile strength, the sample will neck down and then eventually fracture. The term ultimate tensile strength is used for the maximum value of tensile stress that a material can withstand without breaking and is calculated at the maximum tensile force divided by the original cross-sectional area.

Because the onset of deviation from a straight line in the elastic region is often very gradual, the "yield strength" of a metal is defined as the stress at 0.002 permanent (or plastic) strain. Continued plastic flow beyond the elastic limit produces increasing stress levels—a process called work hardening. During this stage, the sample deforms uniformly, elongating and thinning while the volume remains constant, until work hardening can no longer keep up with the continuing increase in stress caused by the reduction in the sample's cross-sectional area. At this point, the stress goes through a maximum, called the ultimate tensile strength, and the sample begins to deform non-uniformly, or neck, before it fractures in a ductile manner. In soft, annealed fcc metals, the typical total plastic (or permanent) strain immediately prior to fracture is 0.20 to 0.50.

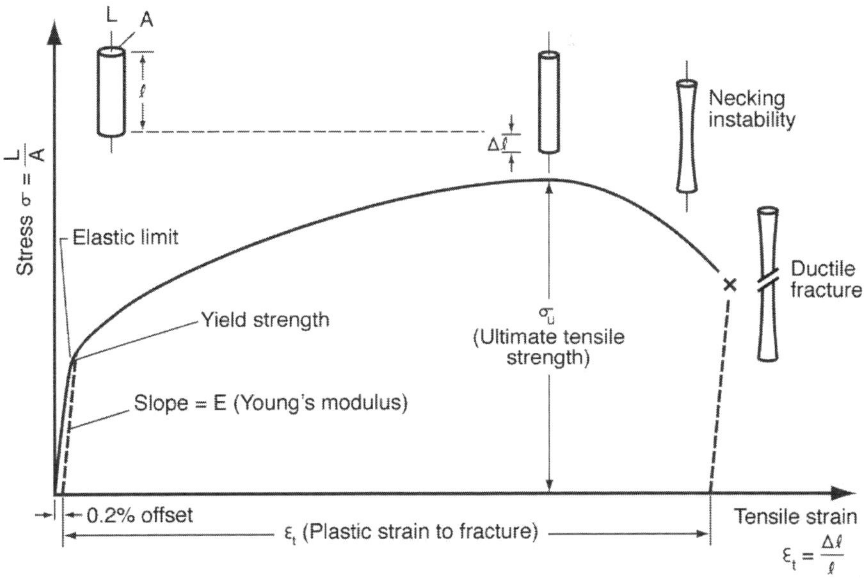

Figure 4. Stress strain graph for a typical metal [4] .

For most engineering purposes, metals are regarded as having failed once they have yielded and are normally loaded at well below the yield point. The calculations in this study show that the cladding probably will exceed the yield point but thermal expansion of the fuel produces very small elongations (strains) from 35°C (95°F) to 600°C (1112°F). So even though the yield point may be exceeded, it is likely that the plastic strain encountered in the cladding will be only a small percent of the failure strain at fuel temperatures below the 640°C (1184°F) fuel melting temperature so that cladding failure is not expected.

Blanford [13] measured the stress strain for 304 L stainless at four temperatures (see Figure 5, note 1 ksi = 1000 psi). At room temperature (RT), the yield strength of the test specimens varied from 37,800 to 44,300 psi. The ultimate strength varied from 95,700 to 97,300 psi for the 304 L. The engineering total strain, or engineering strain at fracture, varied from 71% to 82%. Measured reduction in area was quite high (ranging from 76% to 89%) showing good ductility qualities over the temperature range.

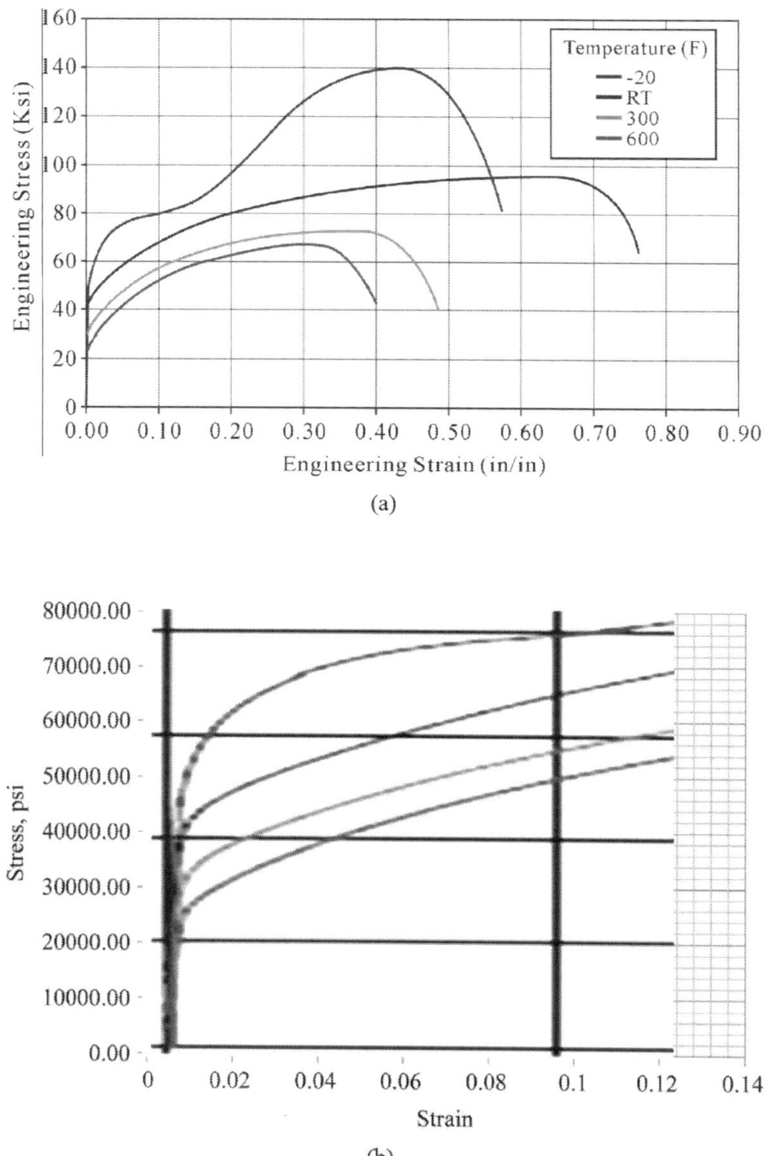

Figure 5. (a) Stress strain measurements based on original cross-sectional area; (b) Data expanded for small strain region (black is room temperature).

Although the stress at different temperatures will be evaluated, the stress strain curve from Figure 5(a) at room temperature will be used in the analysis for all temperatures. The model results indicate that the strain will be greater than the yield strain for stainless steel 304 L but is still very small. The ratio of the elongation/original length for unrestrained expansion of the fuel from room temperature to 600°C (1112°F) using a high value of the coefficient of thermal expansion is 0.0161 in/in. This is the maximum strain the clad could undergo for a fuel element at or less than 600°C (1112°F). This strain only corresponds to about 1/10th of the smallest division (0.1 in/in) in Figure 5(a).

As mentioned, the stress strain curve for 304 L stainless steel was approximated with the room temperature stress strain curve. Only the small strain region was considered since the strains are small. The data for the small strain region is shown in Figure 5(b).

4. ANALYTICAL MODELS

Two models are considered in this paper, the first assumes linear stress strain materials for both the fuel and the cladding, the second allows the cladding to have a non-linear stress strain relationship. The linear model produces very high stresses whereas the non-linear allows the cladding to plastically deform which relieves the high stress without leading to failure of the cladding. Both these models assume there is very little friction between the fuel and the cladding. Calculations show that this is a reasonable assumption for most of the fuel elements.

4.1. Stress Analysis Assuming the Stress Is Below the Yield Stress

This section presents equations used to calculate the compression in the fuel plate and the tension in the cladding for a fuel element raised to a given temperature assuming a linear stress strain material. The assumption is made that the stress is zero at room temperature. Also, due to the large conductivity of both the fuel and the cladding metals, both are probably at the same temperature but different temperatures are assumed just to keep the solution general. The linear model developed to determine this stress is presented in this section. It is realized that the fuel element is at a slightly higher temperature than room temperature when manufactured so that this calculation may be slightly conservative.

The length of the fuel and cladding when the fuel element is raised to a higher temperature is evaluated in a two step process. First, the length of both the fuel and cladding is evaluated at the new temperature as if each were unconstrained. The second step consists of finding the tensile stress in the cladding and the compressive stress in the fuel that will bring them both to the same length.

Step 1 The change in length each material undergoes if each were unconstrained to expand would be:

$$\Delta L_C = \alpha_C L_{C0}\left(T_C - T_0\right) \tag{1}$$

$$\Delta L_F = \alpha_F L_{F0}\left(T_F - T_0\right) \tag{2}$$

where T_0 = Temperature at which the fuel element was made T_F = Fuel temperature at new time, $t = t_1$

T_C = Clad temperature at new time, $t = t_1$

α_i = Coefficient of thermal expansion, CTE The new lengths at the new temperature are obtained by adding the change to the original length:

$$L_{C1} = L_{C0} + \Delta L_C \tag{3}$$

$$L_{F1} = L_{F0} + \Delta L_F \tag{4}$$

Step 2 The fuel will want to expand more than the cladding since $\alpha_F > \alpha_C$. However, the cladding end caps will constrain the fuel to keep the clad and fuel the same length. This requires that the tensile force on the clad, F_C, to be equal to and of opposite sign as the compressive force on the fuel, F_F since equilibrium on the fuel element requires

$$F_C + F_F = 0 \tag{5}$$

The length of the fuel and cladding, L, which satisfies this condition can be obtained by applying Hooke's law to each metal to bring each metal to this unknown length. Then by substituting into Equation (5), the length L can be obtained.

Assuming that both materials are in the linear range, the tensile stress, σ_C, required to lengthen the clad to the common length L using Hooke's law is:

$$\sigma_C = \frac{F_C}{A_C} = \frac{E_C}{1-\epsilon_C}\frac{(L-L_{C1})}{L_{C1}} \tag{6}$$

where A_C = Area of the clad E_C = Young's modulus for the clad

ϵ_C = Poisson's ratio, dimensionless

$(L-L_{C1})/L_{C1}$ = cladding strain, ε_C.

The compressive stress, σ_F, needed to compress the fuel to the common length L is:

$$\sigma_F = \frac{F_F}{A_F} = \frac{E_F}{1-\epsilon_F}\frac{(L-L_{F1})}{L_{F1}} \tag{7}$$

where the same definitions apply to the fuel.

Substituting these two forces into Equation (34), which add up to zero, gives:

$$\frac{E_C}{1-\epsilon_C}\frac{(L-L_{C1})}{L_{C1}}A_C+\frac{E_F}{1-\epsilon_F}\frac{(L-L_{F1})}{L_{F1}}A_F=0$$

or rearranging to collect terms containing L yields

$$\left(\frac{E_C}{1-\epsilon_C}\frac{A_C}{L_{C1}}+\frac{E_F}{1-\epsilon_F}\frac{A_F}{L_{F1}}\right)L-\left(\frac{E_C}{1-\epsilon_C}A_C+\frac{E_F}{1-\epsilon_F}A_F\right)=0$$

This equation may be solved for L:

$$L=\frac{\left(\frac{E_C}{1-\epsilon_C}A_C+\frac{E_F}{1-\epsilon_F}A_F\right)}{\left(\frac{E_C}{1-\epsilon_C}\frac{A_C}{L_{C1}}+\frac{E_F}{1-\epsilon_F}\frac{A_F}{L_{F1}}\right)} \tag{8}$$

This value of L is substituted into Equations (6) and (7) to get the strains and stresses.

$$\varepsilon_C=\frac{(L-L_{C1})}{L_{C1}};\varepsilon_F=\frac{(L-L_{F1})}{L_{F1}};\sigma_C=\frac{E_C}{1-\epsilon_C}\varepsilon_C;\sigma_F=\frac{E_F}{1-\epsilon_F}\varepsilon_F \tag{9}$$

4.2. Stress Analysis with the Stainless Steel Stressed Beyond the Yield Stress

In performing the stress calculations with the above equation, the stress in the clad for higher temperatures exceeds the yield stress for stainless steel. The above analysis was modified to take the non-linear stress strain curve into account.

The force must still be in equilibrium so that Equation (5) is still valid:

$$F_C+F_F=0 \tag{10}$$

The tensile stress, σ_C, required to lengthen the clad to the common length L is:

$$\sigma_C = \frac{F_C}{A_C} = \frac{1}{1-\epsilon_C} E_C\{\varepsilon_C\}; \varepsilon_C = \frac{(L-L_{C1})}{L_{C1}} \tag{11}$$

where $E_C\{\varepsilon_C\}$, a function of the strain, has replaced the product of the Young's Modulus and the cladding strainfor the cladding in Equation (6). The equation which is used to represent the room temperature data in Figure 5(b) is

$$E_C(\varepsilon_C) = E_C\left\{\frac{(L-L_{C1})}{L_{C1}}\right\} = \min(3.62\text{E}+07*\varepsilon_C; 2.42\text{E}+05*\varepsilon_C + 43352.92) \tag{12}$$

As before, the compressive stress, σ_F, needed to compress the fuel to the common length is:

$$\sigma_F = \frac{F_F}{A_F} = \frac{E_F}{1-\epsilon_F}\frac{(L-L_{F1})}{L_{F1}} \tag{13}$$

Substituting the two forces from these stresses into Equation (10) gives:

$$G = \frac{1}{1-\epsilon_C} E_C\left\{\frac{(L-L_{C1})}{L_{C1}}\right\} A_C + \frac{E_F}{1-\epsilon_F}\frac{(L-L_{F1})}{L_{F1}} A_F = 0 \tag{14}$$

Now the value of L is solved for by iteration to make the above function G equal zero.

As before, this value of L is substituted into Equations (39) and (40) to get the strains and stresses.

$$\varepsilon_C = \frac{(L-L_{C1})}{L_{C1}}; \varepsilon_F = \frac{(L-L_{F1})}{L_{F1}}; \sigma_C = \frac{E_C(\varepsilon_C)}{1-\epsilon_C}; \sigma_F = \frac{E_F}{1-\epsilon_F}\varepsilon_F \tag{15}$$

Using the more realistic values of CTEs requires taking into account that the fuel expansion includes step changes in the fuel CTE. This is included in the model by modifying Equation (2) for the fuel length thermal expansion as:

$$\text{for } T_F < T_1 \Delta L_F = \alpha_{F0} L_{F0} (T_F - T_0)$$
$$\text{for } T_F < T_2 \Delta L_F = \alpha_{F0} L_{F0} (T_1 - T_0) + \alpha_{F1} L_{F0} (T_F - T_1)$$
$$\text{for } T_F > T_2 \Delta L_F = \alpha_{F0} L_{F0} (T_1 - T_0) + \alpha_{F1} L_{F0} (T_2 - T_1) + \alpha_{F2} L_{F0} (T_F - T_2) \quad (16)$$

where T_i is the temperature at which the CTE changes from $\alpha_{F,i-1}$ to $\alpha_{F,1}$.

5. RESULTS

A conservative calculation was first performed using a constant CTE for both materials with α_F= 0.000028/°C and α_C= 0.000016/°C. In reality, as pointed out in Section 1.2, the fuel CTE is about the same as the cladding CTE at low temperatures but increases at higher temperatures. A second more realistic calculation was performed with variable values later in this section.

The range of strain which corresponds to the amount of strain calculated for the ZPPR fuel cladding is a small fraction of the total range in Figure 5(a). This smaller range is shown in Figure 6 as well as the approximate fit (Equation (12)) used to represent the data. It is noted that once the cladding yield stress is exceeded, the cladding stress in the plastic region increases only a small amount for large increases in the strain.

5.1. Results with Constant CTE's

The calculated increase in length of the 8-in.-long fuel for the complete range of temperatures of interest is shown in Figure 7. Results from both the elastic (linear) and elastic-plastic (non-linear) models are shown. The lower (red) line is the result obtained assuming a constant Young's modulus (that is, the elastic model), and the upper (blue) is obtained with the non-linear stress strain curve of Figure 6 (that is, the elastic-plastic model). The temperature corresponding to the yield stress temperature is about 170°C (338°F) for this set of CTEs. It is noted that the total elongation is very small (0.05 in.) even for the highest temperature 600°C.

The strain is shown in Figure 8. Note that the strain is less for the elastic model (lower line) than for the elastic-plastic model (upper line). Both models are the same until the yield stress temperature is reached and then they deviate.

The stress is shown in Figure 9. The stress is much higher when using the elastic model (upper line) than the elastic-plastic model (lower line). The elastic-plastic stress appears to be almost constant for fuel temperatures above the yield (even though it is increasing a small amount with increasing temperature). It is seen that a small amount of plastic strain resulting from the elastic-plastic model results in a significant stress relief.

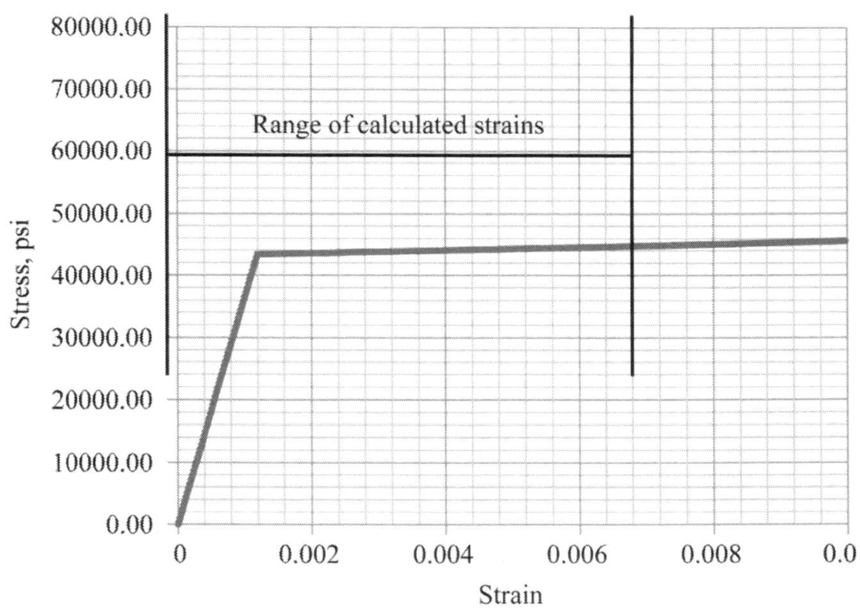

Figure 6. Expanded strain range of Figure 5.

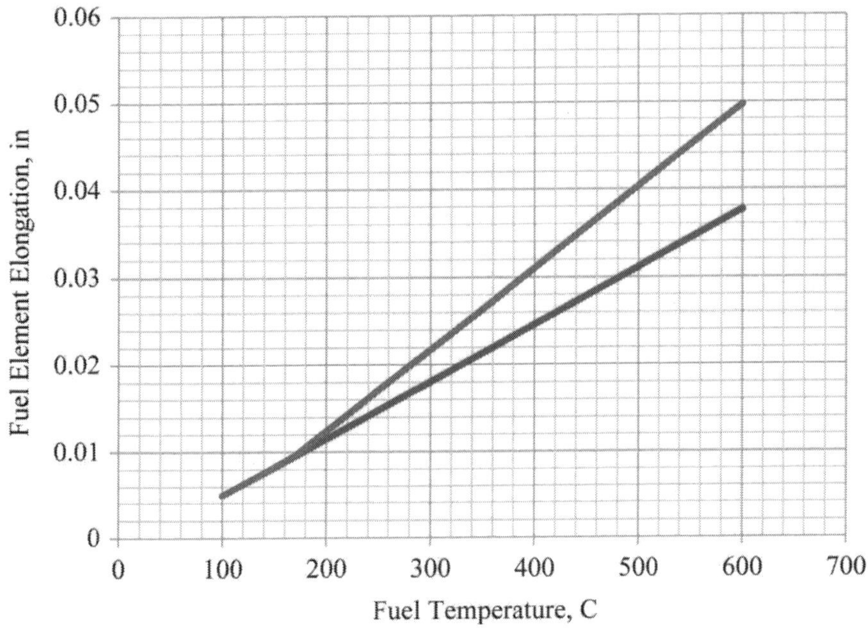

Figure 7. Elongation for conservative thermal expansion coefficients.

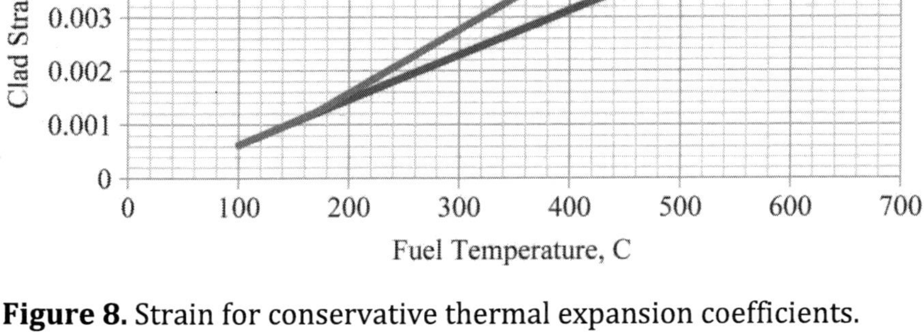

Figure 8. Strain for conservative thermal expansion coefficients.

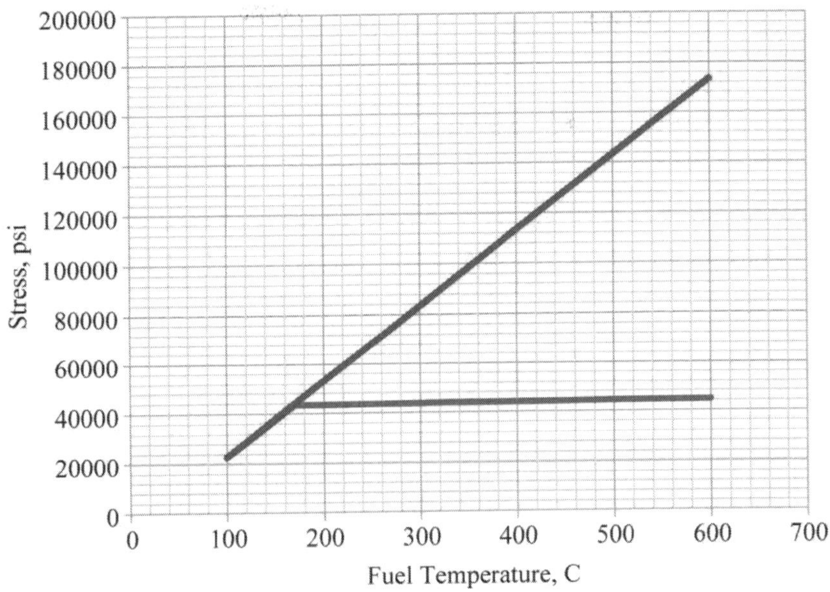

Figure 9. Stress for conservative thermal expansion coefficients.

5.2. Variable CTE Results

The stress and strain calculation was repeated assuming that consistent with Section 1.2.1 the CTE for the fuel increases at higher temperature. The clad remains constant at 0.000016/C. The fuel values are approximated by $T_F < 227$ CTE = 0.000016/C, $227 < T_F < 545$ CTE = 0.000028/C, $545 < T_F$ CTE = 0.000052/C.

The stress plot (lower line) is presented in Figure 10. The original constant CTE elastic model (upper line) is included for comparison.

The clad stress is zero when the CTEs of cladding and fuel are the same up to 227°C. Then the clad stress increases at the same rate as the elastic model until the clad yield point is reached at 370°C (698°F). The maximum elastic-plastic stress reaches approximately the same value as in the previous constant CTE case but does not reach it until 370°C (698°F). The rate of stress increase is very small after this with a slight increase when the CTE of the fuel increases to 0.000052/C at 545°C.

The strain (lower line) is shown in Figure 11. It remains zero until the fuel CTE increases to 0.000028/C. Then it starts to increase with the same slope as the elastic (upper line) model until the elastic limit is reached where the strain increases at a faster rate. At 545°C, the fuel CTE increases and the strain rate increases again to accommodate it even though the stress does not increase much.

6. COMPARISON OF THE CALCULATED STRESS TO THE ULTIMATE STRESS

As previously mentioned, the ultimate stress is defined as the maximum stress at which a material will not rupture. So it is probable that no holes will develop in the cladding if the stress is below this value. The ultimate stress (as well as the yield stress) varies with temperature as is shown in Figure 5(a). The ultimate stresses for three temperatures for which he presents yield stress data (room temp, 149°C [300°F], and 316°C [600°F]) are

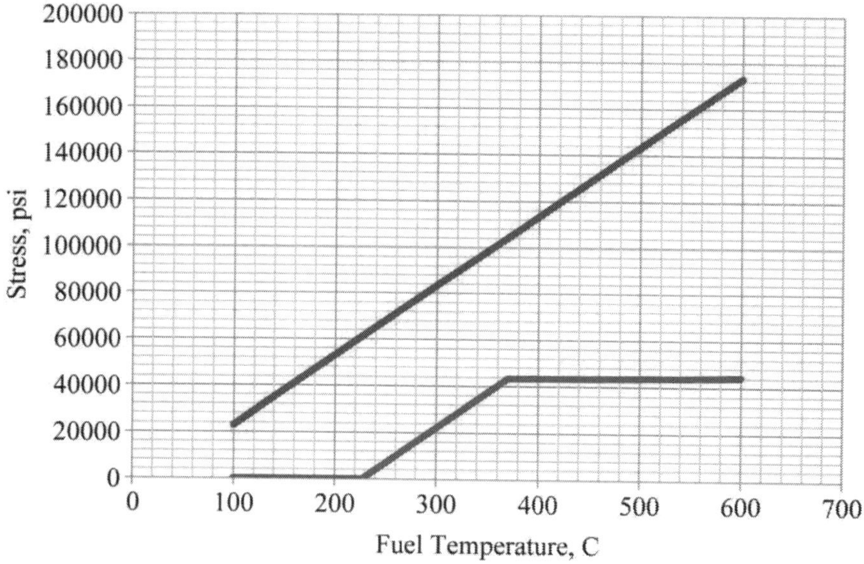

Figure 10. Stress best estimate thermal expansion.

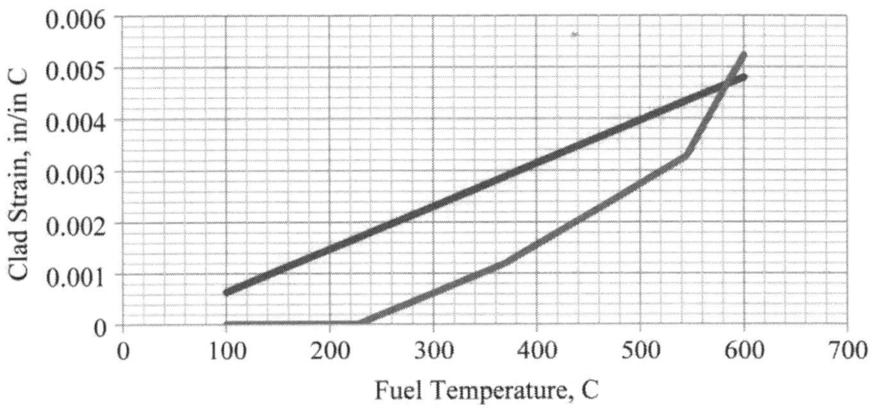

Figure 11. Strain best estimate thermal expansion.

plotted in Figure 12. The data are extrapolated to 600°C (1112°F) with a straight line which is probably conservative since the three data points indicate that the ultimate stress decreases at a decreasing rate with temperature.

A similar plot is included in Figure 13 for the strain. It has also been extrapolated by a straight line which is also probably conservative.

The highest temperature calculated in Ref. [2] is less than 290 (554°F), which, according to Figure 12, has an ultimate stress of over 68,000 psi and, according to Figure 13, an ultimate strain of 0.325. Note, the data covers this maximum calculated temperature. The maximum stress calculated with the elastic plastic model is 42,000 psi at 600°C. The maximum strain calculated is less than 0.005 for the maximum design temperature of 600°C. In soft, annealed fcc metals, the typical total plastic (or permanent) strain immediately prior to fracture is 0.20 to 0.50 (according to Ref. [4]). Therefore, it is concluded that since the calculated stresses are far below the ultimate stress and the maximum strain is far below the 0.20 strain lower limit damage criterion, no failures or holes in the cladding would be expected due to thermal expansion at the storage temperatures of the fuel.

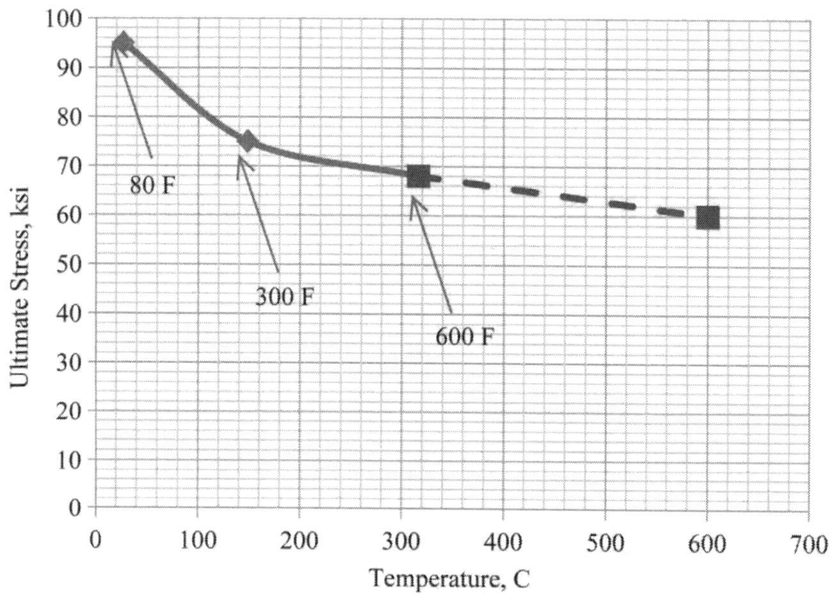

Figure 12. Extrapolated ultimate stresses from Blanford (Ref. Blanford, 2007).

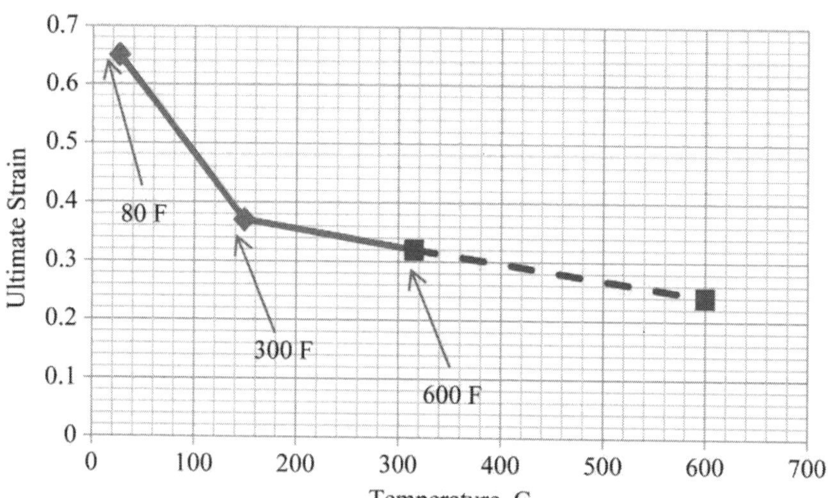

Figure 13. Extrapolated ultimate strains from Blanford [13] .

Two additional items should be noted which reinforce this conclusion. The first is that the analysis was done using room temperature stress strain data. If higher temperature data were used, the calculated stress would be lower since the cladding would start to yield at a lower temperature and lower stress although the strain would be higher. The second point noted is that the stress reached with the realistic thermal expansion coefficient is less than 20,000 psi and the strain is only 0.0005 at 290°C.

7. CONCLUSIONS

The conclusion reached based upon the previous data and analysis is that it is unlikely that the fuel element clad could be compromised at the storage temperatures calculated or in any reasonable temperature excursion in the storage room.

There are two items that could affect the ZPPR fuel integrity which are not accounted for in this paper. The first is the fact that the material yield behavior data is based on stretching metal bars and assuming the same stress strain behavior applies to a thin metal flat plate 15 mils thick.

The second concerns the differential pressure across the cladding. During initial fuel manufacture in an argon atmosphere, the fuel is evacuated to between one half and one third atmosphere prior to closure. After closure when the fuel is exposed to atmospheric pressure, the free volume between cladding and fuel decreases. Calculations indicate that the internal pressure of the fuel increases to atmospheric and that complete cladding collapse does not occur in most cases. This means there will be some cladding bending due to atmospheric pressure changes that occur over the years of storage so that some fatigue failure of fuel is possible.

Helium is generated in the fuel due to the alpha decay of Pu-241 to Am-241. If this helium diffuses out of the metal and is retained in the fuel plenum, it will increase the internal fuel pressure. The fuel with the high concentrations of Pu-241 has the highest decay heat (and consequently the highest temperatures) would generate the most helium. The higher temperature of this fuel will also increase the fuel pressure.

ACKNOWLEDGEMENTS

This project was supported by the US Department of Energy, Science and Technology, under contract W- 31-109-ENG-38. This manuscript has been authored by Battelle Energy Alliance, LLC under Contract No. DE-AC07-05ID14517 with the US Department of Energy. The United States Government retains and the publisher, by accepting the article for publication, acknowledges that the United States Government retains a nonexclusive, paid-up, irrevocable, world-wide license to publish or reproduce the published form of this manuscript, or allow others to do so, for United States Government purposes.

REFERENCES

1. Solbrig, C.W., Krsul, J.R. and Olsen, D.N. (1994) Pyrophoricity of Uranium in Long Term Storage Environments. Proceedings of DOE Spent Nuclear Fuel: Challenges and Initiatives, ANS Topical Meeting, Salt Lake City, 13-16 December 1994, 89-93.

2. Solbrig, C.W., Pope, C.L. and Andrus, J.P. (2013) Analysis of Possible Damage of ZPPR Pu Fuel in Storage. ECAR 2121, Idaho National Laboratory, Idaho Falls.

3. TEV-1473 (2012) Summary of ZPPR Pu Plate and Pin Inspection and Operation Records. Rev. 0.

4. Hecker, S.S. and Stevens, M.F. (2000) Mechanical Behavior of Plutonium and Its Alloys. Los Alamos Science, Los Alamos.

5. Till, C.E., Chang, Y.I. and Hannum, W.H. (1997) The Integral Fast Reactor—An Overview. Progress in Nuclear Energy, 31, 3-11.

6. Johnson, M.L. (1965) The Description of Fuel and Other Core Materials Procured for SEFOR Critical Experiments in ZPR-III. GEAP 5133, AEC Research and Development Report, General Electric Co., San Jose.

7. Steel, A.K. (2007) 304/304l Stainless Steel. Product Data Bulletin.

8. PF-1600 (1967) Specification for U-Pu-Mo Alloy ZPPR Fuel Elements. Drawings on Pages 42 to 45 Inclusive, Argonne National Laboratory, Chicago.

9. Lawson, A.C., et al. (2004) Invar Model for δ-phase Pu: Thermal Expansion, Elastic, and Magnetic Properties. LA UR 04-6008, Los Alamos National Laboratory, Los Alamos.

10. ANL-7471 (1972) Final Safety Analysis Report of the Zero Power Plutonium Facility.

11. Klann, R.T., Austin, B.D., Aumeier, S.E. andOlsen, D.N. (2001) Inventory of Special Nuclear Material from the Zero Power Physics Reactor. Argonne National Laboratory-West.

12. Anonymous (2013) Engineering Toolbox. http://www.engineeringtoolbox.com/surface-roughness-ventilation-ducts-d_209.html

13. Blandford, R.K., Morton, D.K., Snow, S.D. and Rahl, T.E. (2007) Tensile Stress-Strain Results for 304L and 316L Stainless Steel Plate at Temperature. INL/CON-06-11962, ASME Pressure Vessels and Piping Division Conference, July 2007, 1-12.

4

CHAPTER

HEAT TRANSFER CHARACTERISTICS OF SQUARE MICRO PIN FINS UNDER NATURAL CONVECTION

Naoko Matsumoto, Toshio Tomimura, Yasushi Koito

Department of Advanced Mechanical Systems, Kumamoto University, Kumamoto, Japan

ABSTRACT

In order to comply with the recent demand for downsizing of the electric equipment, the minia- turization and the improvement in heat transfer performance of a heat sink under natural air-cooling are increasingly required. This paper describes the experimental and numerical investigations of heat sinks with miniature/micro pins and the effect of the pin size, pin height and the number of pins on heat transfer characteristics of heat sinks. Five types of basic heat sink models are investigated in this study. The whole heat transfer area of heat sinks having the different pin size, pin height and the number of pins respectively is kept constant. From a series of experiments and numerical analyses, it has been clarified that the heat sink temperature rises with increase in the number of pins. That is, the heat sink with miniaturized fine pins showed almost no effect on the heat transfer enhancement. This is because of the choking phenomenon occurred in the air space among the pin fins. Reflecting these results, it is confirmed that the heat transfer coefficient reduces with miniaturization of pins. Concerning the effects of the heat transfer area on the heat sink performance, almost the same tendency has been observed in other three series of large surface area, that is, higher pin height. Furthermore as a result of studying non-dimensional convection heat transfer performance, it was found that the relation between the Nusselt number (Nu) and the Rayleight number (Ra) is given by $Nu = 0.16\ Ra^{0.52}$.

Keywords: Natural Air-Cooling, Heat Sink, Micro Pin Fin, Heat Transfer Performance, Experiment, Numerical Analysis

1. INTRODUCTION

In electronic equipment applications with high heat fluxes, the heat transfer performance and the lifetime of components are often deteriorated due to temperature rise. To avoid those unfavorable problems, it is necessary to keep the component under a critical temperature by introducing a high performance heat sink. Furthermore, in order to comply with the recent demand for downsizing of the electronic equipment, the miniaturization and the improvement in heat transfer performance of the heat sink are required much more strongly than ever. But it is well known that the decrease in heat transfer area by downsizing leads to a decline of the heat sink performance.

Many studies of heat transfer characteristics using large scaled heat sinks have been carried out so far. For example, Aihara et al. [1] and Sara [2] revealed that the heat sink performance depends on the pin size, the population density of pin. Zografos and Sunderland [3] studied the effect of pin arrangement on the heat transfer performance, and reported that a choking phenomenon, which had a negative effect on heat transfer from a heat sink, was observed in the air space among pin fins. Huang et al. [4] , Sparrow et al. [5] and Sertkaya et al. [6] investigated experimentally the dependence of the pin fin performance on a heat sink orientation.

In electronic equipment applications, heat sinks are placed within enclosed areas and the emphasis is on making the enclosures more compact than ever. Yu et al. [7] and Bocu et al. [8] reported natural convection heat transfer from heat sink attached to enclosures in their studies.

On the other hand, concerning small sized heat sinks or nano particles structure, the analytical and experimental studies were published by Narasimhan et al. [9] and Minakami et al. [10] , Kunugi et al. [11] . Although interesting findings on natural and forced convection, and radiation from heat sinks have been reported in the previous studies, the mechanism of the heat transfer has not been clarified yet.

Accordingly, in the present study, the effects of pin size, pin height and the number of pins on the heat transfer characteristics of heat sinks with miniature/micro pins have been investigated fundamentally.

2. HEAT SINK MODELS

Five types of basic heat sink models are shown in Figure 1. The first model has four square pins, and is named as Type 1. Each pin placed at equal space of the heat sink base (length: 25 mm, width: 25 mm, thickness: 2 mm) has the pin width w_p = 6.25 mm and the pin height h_p = 6.5, 14, 21.5, 29 mm at intervals of 7.5 mm. And in Type 2 model, the pin size is reduced to half the pin width and height of Type 1. That is, under the conditions of constant heat transfer area, the number of pins for Type 2 is quadrupled against Type 1. In the same manner, five types of heat sinks in total, which have the pin height from several hundred micro meters to a few millimetres and have the same heat transfer area, are prepared.

As listed in Table 1, the number of pins N is changed from 4 to 1024, and corresponding to this change in N, the pin width w_p and the pin height h_p vary from 0.39 mm to 6.25 mm and from 0.41 mm to 29 mm, respectively. The whole heat transfer area of the square pin series A_{hs} is 1.48 ´ 10^{-3} m^2, 2.23 ´ 10^{-3} m^2, 2.98 ´ 10^{-3} m^2 and 3.73 ´ 10^{-3} m^2.

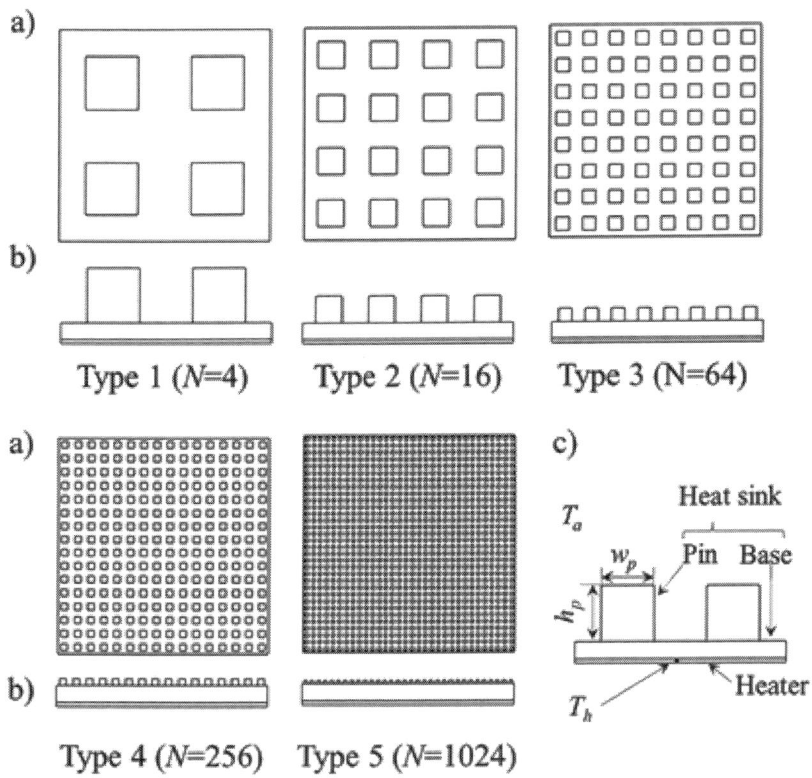

Figure 1. Heat sink models, (a) Top view, (b) Side view, (c) Parts and nomenclature.

Table 1. Dimensions of pin fin.

Type	N (−)	w_p (mm)	h_p (mm)			
			●	□	◆	△
1	4	6.25	6.5	14	21.5	29
2	16	3.13	3.25	7	10.75	14.5
3	64	1.56	1.63	3.5	5.38	7.25
4	256	0.78	0.81	1.75	2.69	3.63
5	1024	0.39	0.41	0.88	1.34	1.81

3. EXPERIMENTAL

Figure 2 shows a schematic diagram of experimental apparatus, which consists of a chamber, a vacuum pump, a vacuum gauge, and a heat sink with a heater used as the heat source. The chamber made of SUS is 400 mm in diameter and 170 mm in height. Here, its wall temperature which is regarded as the ambient temperature T_a was kept at 25°C by circulating cooling water. The heat sink was hung in the air in the chamber by using fine wires to reduce the heat loss from the heat sink to the chamber wall by conduction. The heat sink was placed in the upward facings of the heated surface. The pressure in the chamber was controlled with the vacuum pump from the atmospheric pressure to the vacuum condition of approximately 3.5×10^{-2} Pa, and the chamber pressure was measured using the vacuum gauge. Under the atmospheric pressure, heat is transferred from the heat sink to the surroundings by convection and thermal radiation. On the other hand, under the vacuum condition of approximately 3.5×10^{-2} Pa, heat is transferred from the heat sink to the surrounding wall only by thermal radiation.

As shown in Figure 3, the heat sink, which is made of Aluminum alloy A6063 (the thermal conductivity l = 200 W/(m·K), the emissivity e = 0.89) with a base plate of 25 mm long, 25 mm wide and 2 mm thick, was placed on the same sized square copper block (thickness: 2 mm, l = 398 W/(m·K)) through a thermal interface material (thickness: 0.5 mm, l = 3.1 W/(m·K)). Furthermore, on the bottom part of the copper block, a ceramic heater is attached through the thermal interface material. The representative temperature T_h was measured using a f 0.5 mm K-type thermocouple fixed in a small groove machined at the copper block center. In Figure 3, T_a is the ambient temperature.

First, under the atmospheric pressure, the heat rate Q_{in} = 1.25 W was applied to the heat sink by the ceramic heater. Second, the representative temperature T_h was measured under the steady-state condition. Third, under the vacuum condition of approximately 3.5×10^{-2} Pa, some amount of heat was applied to the heat sink by the heater until the heat sink temperature became the same temperature T_h under the atmospheric pressure, and then the heat transferred by thermal radiation Q_r from the heat sink under the steady-state conditions was measured based on the input power to the heater. Furthermore, in the same manner, T_h and Q_r for the heat sink consisting only of the base plate were also measured for reference.

4. NUMERICAL ANALYSIS

Figure 4 shows the boundary conditions for the large air space surrounding the heat sink placed in the upward facings of the heated surface. A series of numerical analyses have been performed for the physical models shown in Figure 1 and for the heat sink consisting only of the base plate (25 mm, 50 mm, 100 mm and 150 mm on a side) for reference. In this study, the fine wires for hanging the heat sink and the copper block, the thermal interface material were neglected for simplification, and only the heater (length: 25 mm, width: 25 mm, thickness: 0.5 mm), the heat sink, and the surrounding large air space were taken into consideration.

The present numerical calculations were carried out using the simulation software of Computational Fluid Dynamics (CFD). The following assumptions are applied in the analysis.

1. Fluid is incompressible.
2. Thermophysical properties are assumed to be constant except density (Boussinesq approximation).
3. Ambient temperature T_a is constant.

Based on these assumptions, the following basic equations are obtained under the steady state conditions.

Continuity equation:

$$\frac{\partial u}{\partial x}+\frac{\partial v}{\partial y}+\frac{\partial w}{\partial z}=0 \tag{1}$$

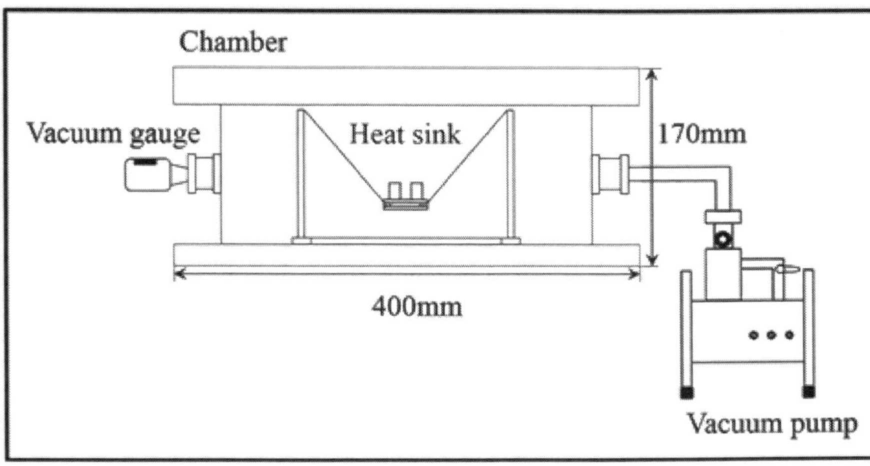

Figure 2. Schematic of experimental apparatus.

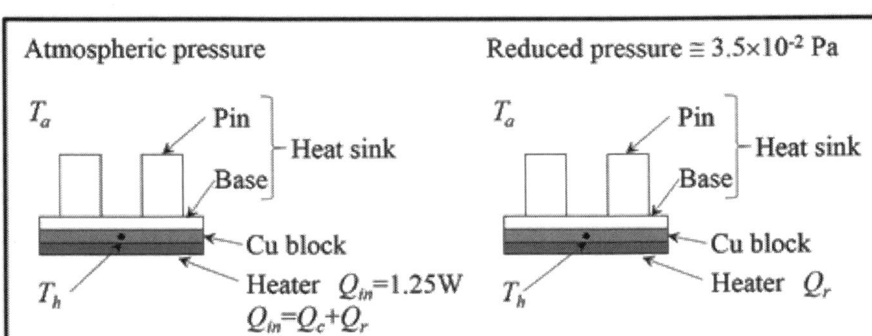

Figure 3. Configuration of tested heat sink.

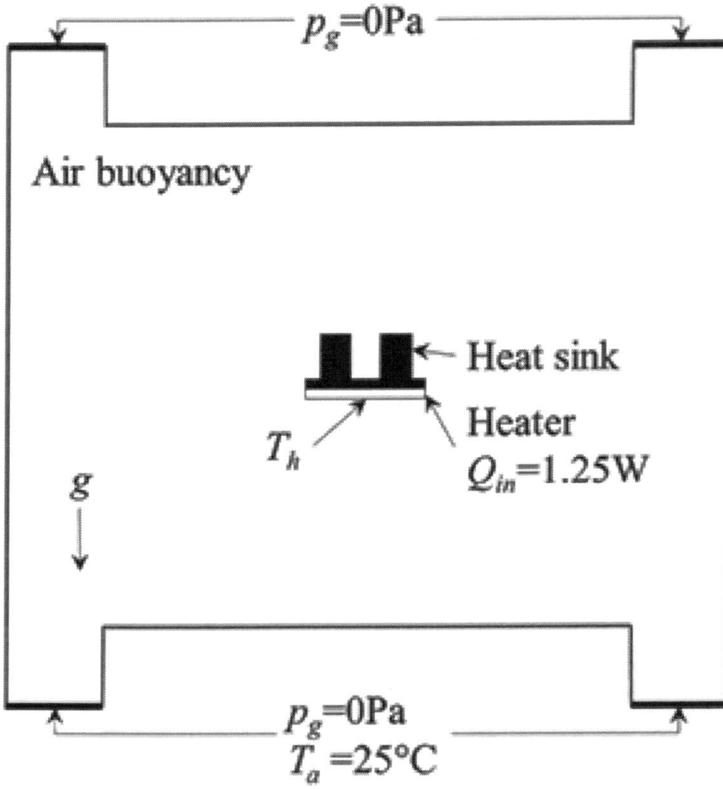

Figure 4. Boundary conditions of CFD (upward facing).

Momentum equations:

$$u\frac{\partial u}{\partial x}+v\frac{\partial u}{\partial y}+w\frac{\partial u}{\partial z}=-\frac{1}{\rho}\frac{\partial p}{\partial x}+\upsilon\left(\frac{\partial^2 u}{\partial x^2}+\frac{\partial^2 u}{\partial y^2}+\frac{\partial^2 u}{\partial z^2}\right) \quad \text{(x-direction)} \quad (2)$$

$$u\frac{\partial v}{\partial x}+v\frac{\partial v}{\partial y}+w\frac{\partial v}{\partial z}=-\frac{1}{\rho}\frac{\partial p}{\partial y}+\upsilon\left(\frac{\partial^2 v}{\partial x^2}+\frac{\partial^2 v}{\partial y^2}+\frac{\partial^2 v}{\partial z^2}\right) \quad \text{(y-direction)} \quad (3)$$

$$u\frac{\partial w}{\partial x}+v\frac{\partial w}{\partial y}+w\frac{\partial w}{\partial z}=-\frac{1}{\rho}\frac{\partial p}{\partial z}+\upsilon\left(\frac{\partial^2 w}{\partial x^2}+\frac{\partial^2 w}{\partial y^2}+\frac{\partial^2 w}{\partial z^2}\right)+g\beta(T-T_a)$$

(z-direction)

(4)

Energy equation:

$$u\frac{\partial T}{\partial x}+v\frac{\partial T}{\partial y}+w\frac{\partial T}{\partial z}=\alpha\left(\frac{\partial^2 T}{\partial x^2}+\frac{\partial^2 T}{\partial y^2}+\frac{\partial^2 T}{\partial z^2}\right)$$

(5)

Boundary conditions are given as follows. The same heat rate Q_{in} = 1.25 W as the experiment was applied to the heat sink. The velocity is zero on all the solid surfaces, that is u = v = w = 0. The gauge pressure p_g of the inlet and the outlet of the large surrounding space is 0 Pa, the inlet air temperature is 25°C, and other remaining surfaces are adiabatic. The amount of heat transfer rate by convection and thermal radiation is equal to that by conduction through the solid.

The equation of conduction within the solid is given by the following equation.

$$\frac{\partial^2 T}{\partial x^2}+\frac{\partial^2 T}{\partial y^2}+\frac{\partial^2 T}{\partial z^2}=0$$

(6)

In the numerical analysis, the same heat rate Q_{in} = 1.25 W as the experiment was applied to the heater. In the case of large sized heat sink consisting only of the base plate, the heat rate added to the base plate was kept constant.

Then the representative temperature T_h of the heat sink under the steady-state conditions and the heat rate released from the heat sink by thermal radiation Q_r were calculated using the simulation software of CFD. In addition, the velocity vectors and the temperature fields at the central lateral plane of the heat sink were obtained.

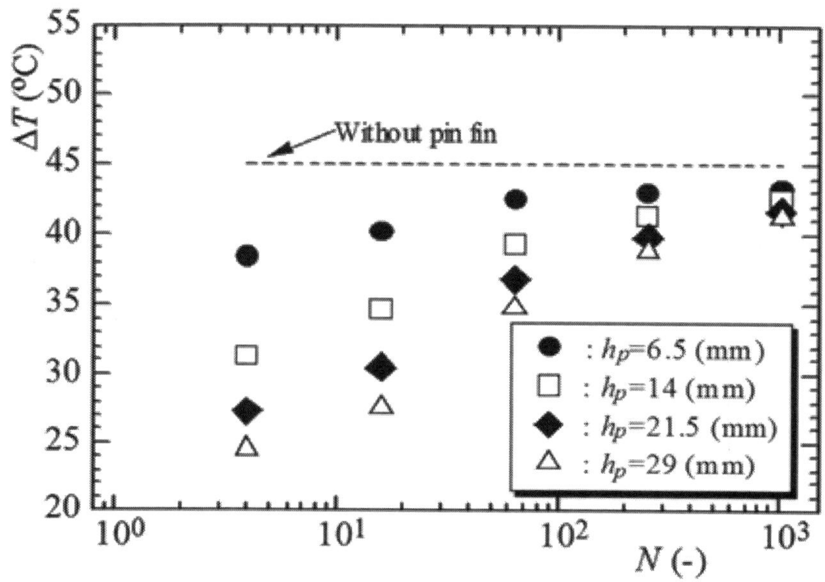

Figure 5. Effects of N and h_p on DT (measured results).

5. RESULTS AND DISCUSSION

5.1. Heat Transfer Performance of Heat Sinks

Figure 5 and Figure 6 show the measured and simulated results on the effects of the number of pins N, the pin height h_p on the temperature rise DT of the heat sink. Here, DT is defined as the difference between the repre sentative temperature T_h and the ambient temperature T_a. Accordingly, the lower DT means a higher heat transfer performance. The symbols show the results for the heat sink with different pin height, that is, heat transfer area. The broken line shows the DT for the heat sink consisting only of the base plate. Further, the number of pins N corresponds to the heat sink type from Type 1 to Type 5.

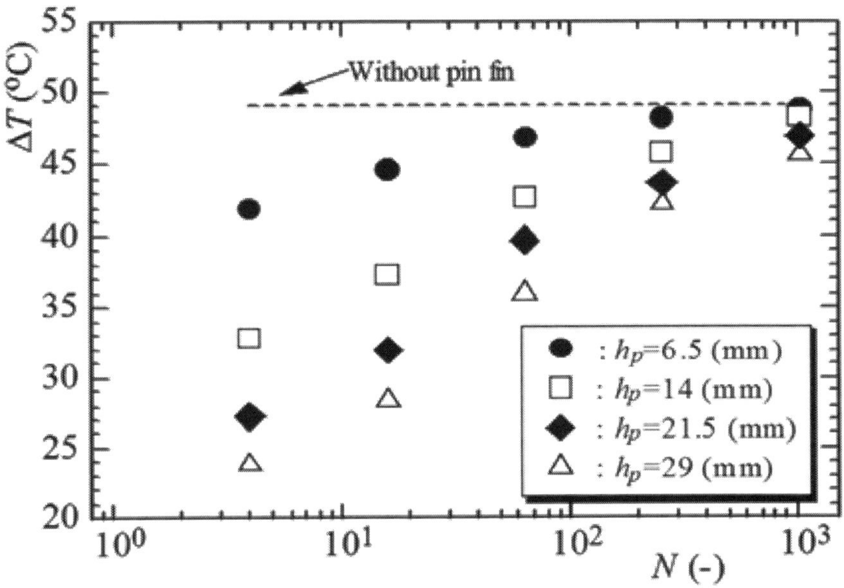

Figure 6. Effects of N and h_p on DT (calculated results).

Concerning Type 1 model having the smallest number of pins, it is found that the heat sink performance of Type 1 model is the highest and as the pin height h_p is higher, the heat sink performance is better. In a series of heat sinks, the heat transfer rate increases with the pin height of Type 1. Each representative temperature T_h of the heat sink rises with increase in the number of pins N. Especially, in the case of large N, the temperature rise DT of the heat sink comes close to that of the base plate level. In other words, the heat sink with miniaturized pins has almost no effect on the heat transfer enhancement. As seen from the figure, the simulated results agree qualitatively well with the measured results. That is, the dependence of the temperature rise DT on the number of pins N agrees fairly well with each other. Unfortunately, however, a quantitative agreement is not obtained between them. This is because the present physical model cannot reproduce the experimental heat sink system exactly.

5.2. Velocity Vector and Temperature Field

The velocity vector in air space surrounding the heat sink model at central longitudinal section, which are obtained from numerical simulation, are shown in Figure 7. From this figure, it can be seen that the velocity vector among pins becomes small with increase in the number of pins. In particular, the air velocity of Type 5 models becomes almost zero.

The temperature fields of the heat sink and its surrounding air space are shown in Figure 8. The representative temperature T_h at the heater surface is the highest. From this figure, it can be seen that the heater tem- perature rises from Type 1 to Type 5. Namely, this means that the increase in the number of pins causes the deterioration of the heat sink performance.

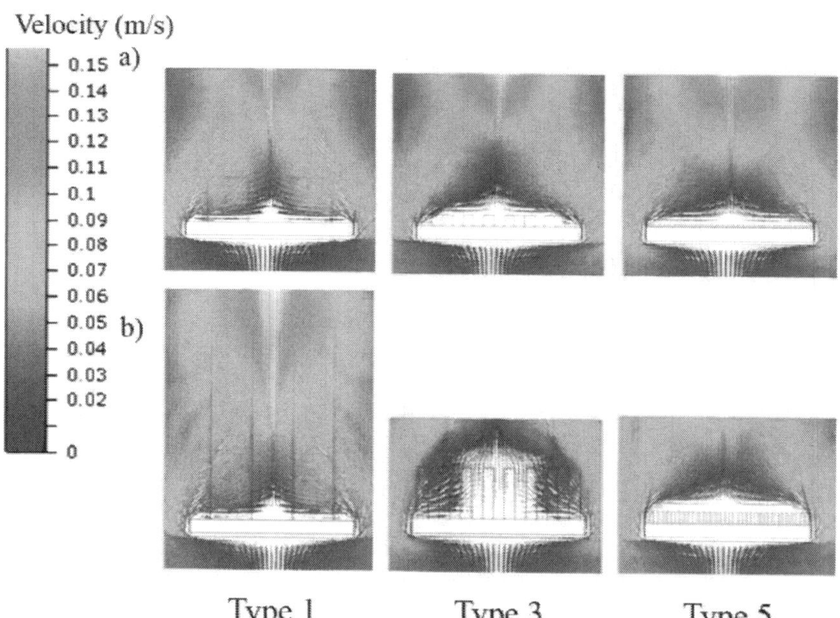

Figure 7. Flow field along center line of heat sink; (a) h_p = 6.5 mm, (b) h_p = 29 mm series.

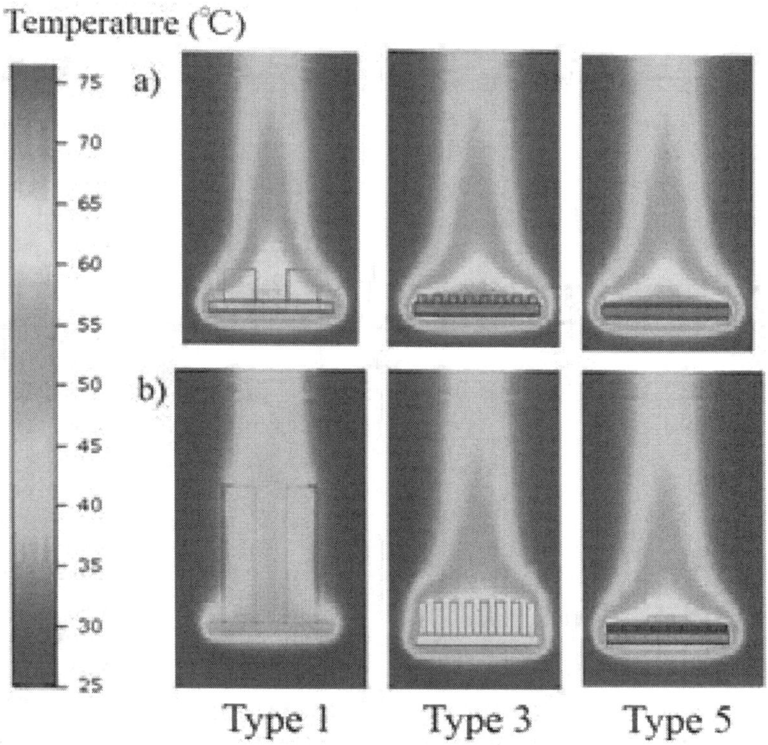

Figure 8. Temperature field along center line of heat sink; (a) h_p = 6.5 mm, (b) h_p = 29 mm series.

5.3. Air Velocity between Pins

Figure 9 shows the air velocity between pins which was numerically obtained for Type 1 to Type 5 models. The present air velocity means the average value obtained at central lateral plane of heat sink. As shown in the figure, the air velocity decreases with increase in the number of pins. This result shows the same tendency as the velocity vectors. In Type 3 to Type 5 models, regardless of the pin height, the air velocity becomes almost zero. This behavior must be the choking phenomena observed in the previous study using the large sized heat sinks done by Zografos et al. [3] , and as an inevitable consequence, it leads to low heat transfer performance of the heat sink.

5.4. The Ratio of Convective and Radiative Heat Transfer Rates

In order to estimate the heat sink performance from the view point of the heat transfer, the ratio of convective and radiative heat transfer rates in whole heat sink system has been also investigated experimentally and analytically, and the measured and calculated results are shown in Figure 10 and Figure 11 respectively. The total amount of heat transfer rate into and out has to keep the energy balance expressed by Equation (7). Under the conditions of the experiments and numerical calculations, Q_{in} is 1.25 W. In addition, the output energy from heat sink is the sum of convective and radiative heat transfer rates Q_c and Q_r, so the amount of heat transferred by convection Q_c is given by the following Equation (8).

$$Q_{in} = Q_{out} \tag{7}$$

$$Q_c = Q_{in} - Q_r \tag{8}$$

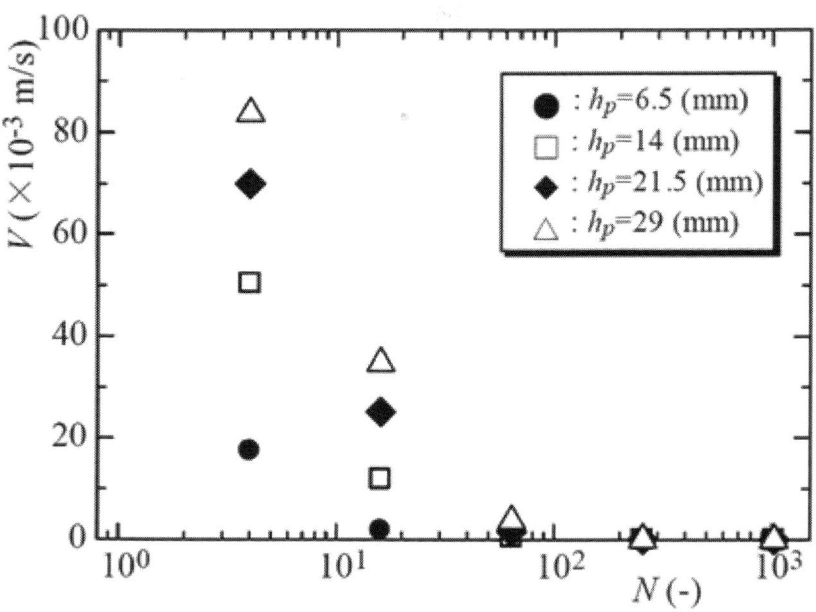

Figure 9. Flow velocity between pins on heat sink.

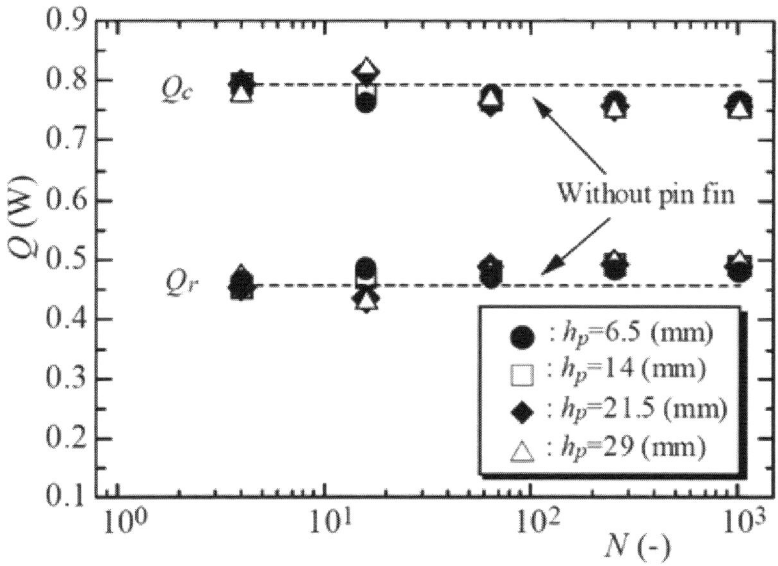

Figure 10. Effects of N and h_p on Q (measured results).

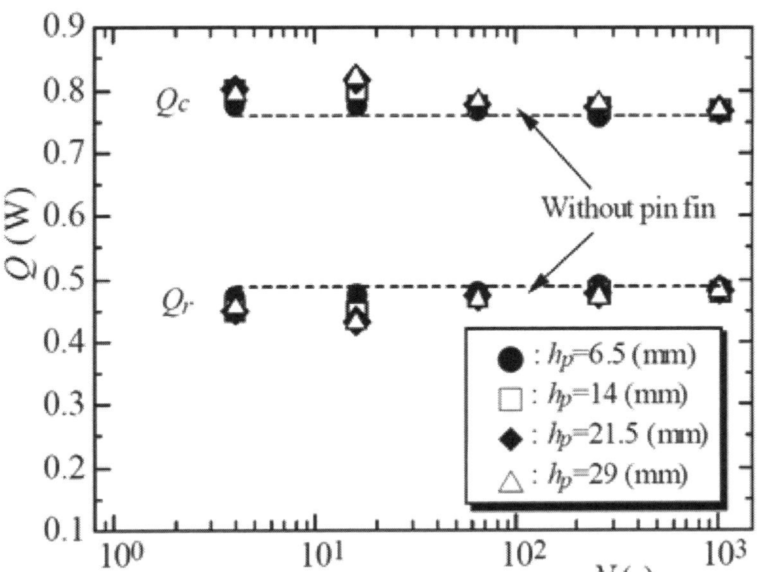

Figure 11. Effects of N and h_p on Q (calculated results).

As seen from Figure 10, it is clear that the convective component Q_c which is the dominant parameter in this case attains its maximum by Type 2, and reduces to Type 5 and then becomes saturated. It can be observed that the convection component Q_c of Type 2 is exactly proportional to the pin height h_p, in other words the heat rate transferred by thermal radiation Q_r vary inversely proportional to the pin height h_p. And in other heat sink models there is no correlation between the convection component Q_c and the pin height h_p. Furthermore, the heat rate transferred by convection Q_c occupies approximately 60% of the total input or output energy Q_{in} or Q_{out}, the heat rate transferred by thermal radiation Q_r is about 40%, and the result agrees very well with the results reported by Sparrow et al. [5] . And from Figure 11, it is found that the calculated results agree well with the measured results qualitatively. However a quantitative agreement is not obtained between them, as previously noted, this is because the present physical model cannot reproduce the experimental heat sink system exactly.

5.5. Heat Transfer of Heat Sink under Natural Convection

In numerical simulations, the total amount of heat transfer rates of the heater and the heat sink was investigated. The convective component of heat sink only is represented by Equation (9) using computable parameters of the total amount of convective heat transfer rate Q_c and the heat transfer from heater by convection Q_{c_ht}.

$$Q_{c\_hs} = Q_c - Q_{c\_ht} \tag{9}$$

The convective heat transfer rate of the heater is calculated by Equation (10),

$$Q_c = \frac{Nu\lambda_a}{L} A(T_h - T_a) \tag{10}$$

$$h = \frac{Nu\lambda_a}{L} \tag{11}$$

where Nu is the Nusselt number, l_a is thermal conductivity of air, L is characteristic dimension and A is heat transfer surface of the heater. Equation (11) gives the heat transfer coefficient h.

It is found that the heater has isothermal heating surface according to the simulated results, the Nusselt numbers at the lower and side surfaces of the heater are calculated by previously-proposed Equation (12) [12] , Equation (13) [13] and Equation (14), where Nu_1 is the Nusselt number of lower surface of heated plate, Nu_2 is that of vertical surface of heated plate. And Ra is the Rayleigh number which is represented as the correlation parameter between the Grashof number and the Prandtl number.

$$Nu_1 = 0.27Ra^{0.25} \tag{12}$$

$$Nu_2 = 0.59Ra^{0.25} \tag{13}$$

where

$$Ra = GrPr = \frac{g\beta(T_h - T_a)L^3}{\upsilon}\frac{\upsilon\rho c_p}{\lambda_a} \tag{14}$$

Two types of characteristic lengths L will be employed here. In the case of the lower surface of cooled plate, L is taken as L = A/P where A is the plate surface area and P is the perimeter which surrounds the area [14] . In the vertical flat plate, L is taken as 0.5 mm which is the height of heater. Using these numbers, Q_{c_ht} of the each side is calculated by substituting Nu_1 and Nu_2 into Equation (10). And then Q_{c_hs} is determined by Equation (9) using known Q_c.

Figure 12 shows the effects of the number of pins and the pin height on the convective heat transfer coefficient. Here, in order to evaluate the improvement in the convective heat transfer performance for heat sink base surface, the surface area of the top of the base is adopted as A (6.25×10^{-4} m^2 constant). In this figure, it is clear that each heat transfer coefficient h of the heat sink decreases with increase in the number of pins N and in the case of Type 5 model it comes close to that of the base plate level. Since Figure 12 agrees well with the results of the temperature rise, it is considered that the factor of decreasing heat transfer performance due to the miniaturization is the convective heat transfer coefficient.

Finally, the calculated data of heat sink, expressed in terms of Nu and Ra, are plotted in Figure 13. In this figure, the lower solid line represents Equation (15) [14] in which Nu_3 is the Nusselt number of upper surface of heated plate and the symbols (′) show the calculated results of the base plate.

$$Nu_3 = 0.54Ra^{0.25} \quad \left(10^4 < Ra < 10^7\right) \tag{15}$$

From these results, it is confirmed that Equation (15) applied to a range of the low-Rayleigh number. Using Equation (11), the Nusselt number of the heat sinks was calculated. Here, L was estimated to be the total value of the pin height h_p and the base height 2 mm. As seen this figure, regardless of the pin size, the pin height, the number of pins, it is revealed that Nu and Ra of the heat sink are expressed by an approximate equation, such as Equation (16).

$$Nu = 0.16Ra^{0.52} \tag{16}$$

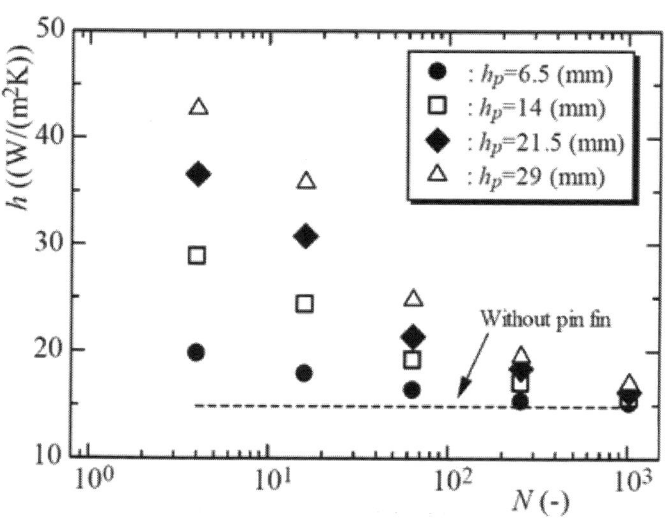

Figure 12. Effects of N and h_p on h (calculated results).

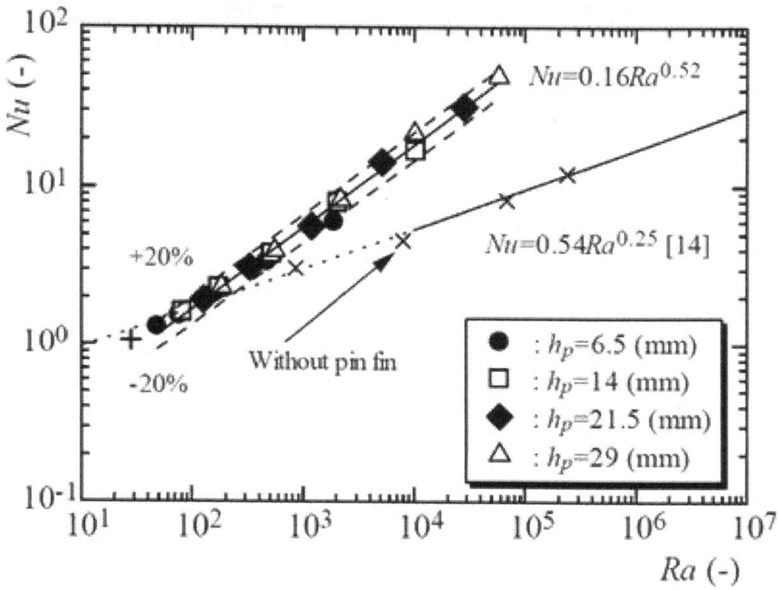

Figure 13. Correlation of Ra and Nu of the heat sink based on the base plate.

Since the calculated results were shown between ±20% of the approximate expression and with miniaturization of pins the Nusselt number Nu of the heat sink comes close to that of the base plate level, it is considered that the proposed correlation implies high accuracy.

6. CONCLUSIONS

In this study, experimental and numerical studies have been done on the effects of the pin size, pin height, the number of pins on the heat transfer performance of the heat sink.

Concerning the heat sinks with the same heat transfer area, it has been confirmed that the heat sink performance changes depending on the population density of pin and the pin size. From the measured and calculated results, the heat sink temperature has been shown to rise with increase in the number of pins. Especially, the heat sink with miniaturized pins has almost no effect on the heat transfer enhancement. It is considered that these characteristics of the heat sink cause the decrease in the convective component, which occupies approximately 60% of total heat transfer rate, and the heat transfer coefficient due to the decrease in the flow velocity when the pin to pin clearances are reduced.

Regardless of the pin size, the pin height, and the number of pins, it has been revealed that the Nusselt number and the Reyleigh number of the heat sink are represented by an approximate equation.

REFERENCES

1. Aihara, T., Maruyama, S. and Kobayakawa, S. (1990) Free Convective/Radiative Heat Transfer from Pin Fin Arrays with a Vertical Base Plate (General Representation of Heat Transfer Performance). International Journal of Heat Mass Transfer, 33, 1223-1232.
2. Sara, O.N. (2003) Performance Analysis of Rectangular Ducts with Staggered Square Pin Fins. Energy Conversion and Management, 44, 1787-1803.
3. Zografos, A.I. and Sunderland, J.E. (1990) Natural Convection from Pin Fin Arrays. Experimental Thermal and Fluid Science, 3, 440-449.
4. Huang, R.T., Sheu, W.J. and Wang, C.C. (2008) Orientation Effect on Natural Convective Performance of Square Pin Fin Heat Sinks. International Journal of Heat and Mass Transfer, 51, 2368-2376.
5. Sparrow, E.M. and Vemuri, S.B. (1986) Orientation Effects on Natural Convection/Radiation Heat Transfer from Pin- Fin Arrays. International Journal of Heat and Mass Transfer, 29, 359-368.

6. Sertkaya, A.A., Bilir, S. and Kargici, S. (2011) Experimental Investigation of the Effects of Orientation Angle on Heat Transfer Performance of Pin-Finned Surfaces in Natural Convection. Energy, 36, 1513-1517.

7. Yu, E. and Joshi, Y. (2002) Heat Transfer Enhancement from Enclosed Discrete Components Using Pin-Fin Heat Sinks. International Journal of Heat and Mass Transfer, 45, 4957-4966.

8. Bocu, Z. and Altac, Z. (2011) Laminar Natural Convection Heat Transfer and Air Flow in Three-Dimensional Rectangular Enclosures with Pin Arrays Attached to Hot Wall. Applied Thermal Engineering, 31, 3189-3195.

9. Narasimhan, S. and Majdalani, J. (2002) Characterization of Compact Heat Sink Models in Natural Convection. IEEE Transactions on Components and Packaging Technologies, 25, 78-87.

10. Minakami, K., Mochisuki, S., Murata, A., Yagi, Y. and Iwasaki, H. (1993) Heat Transfer Characteristics of Pin-Fins with In-Line Arrangement (1st Report, Effect of the Pin Pitch). Japan Society of Mechanical Engineers, 59, 300-307.

11. Kunugi, T., Muko, K. and Shibahara, M. (2004) Ultrahigh Heat Transfer Enhancement Using Nano-Porous Layer. Superlattices and Microstructures, 35, 531-542.

12. Kreith, F. and Bohn, M.S. (2000) Principles of Heat Transfer. 6th Edition, Brooks/Cole.

13. Nakayama, A. (2008) Problems in Heat Transfer. Japan Society of Mechanical Engineers. Maruzen Publishing Co., Ltd.

14. Goldstein, R.J. and Sparrow, E.M. (1973) Natural Convection Mass Transfer Adjacent to Horizontal Plates. International Journal of Heat and Mass Transfer, 16, 1025-1035.

5

CHAPTER

THERMAL STRESS IN HFEF HOT CELL WINDOWS DUE TO AN IN-CELL METAL FIRE

Charles W. Solbrig*, **Stephen A. Warmann**

Idaho National Laboratory, Idaho Falls, USA

ABSTRACT

This work investigates an accident during the pyrochemical extraction of Uranium and Plutonium from PWR spent fuel in an argon atmosphere hot cell. In the accident, the heavy metals (U and Pu) being extracted are accidently exposed to air from a leaky instrument penetration which goes through the cell walls. The extracted pin size pieces of U and Pu metal readily burn when exposed to air. Technicians perform the electrochemical extraction using manipulators through a 4 foot thick hot cell concrete wall which protects them from the radioactivity of the spent fuel. Four foot thick windows placed in the wall allow the technicians to visually control the manipulators. These windows would be exposed to the heat of the metal fire. This analysis determines if the thermal stress caused by the fire would crack the windows and if the heat would degrade the window seals allowing radioactivity to escape from the cell.

Keywords: Pyrophoric Metals, Thermal stress, Nuclear Fuel reprocessing, Hot Cells, brittle Materials, Yield stress

1. INTRODUCTION

This is the first time the complete process of reprocessing PWR fuel to produce fuel for use in Fast Reactors will be demonstrated on a lab scale. This project has introduced a new safety concern in the Hot Fuel Examination Facility hot cell. The concern is the heat generated by the accidental burning of the heavy metals (U and Pu) being extracted when accidently exposed to air from a leaky instrument or manipulator penetration which goes through the cell walls. The extracted pin size pieces of U and Pu metal readily burn when exposed to air. Technicians perform the electrochemical extraction using manipulators through a 4 foot thick hot cell concrete wall which protects them from the radioactivity of the spent fuel. Four foot thick windows placed in the wall allow the technicians to visually control the manipulators. These windows would be exposed to the heat of the metal fire. This analysis determines if the thermal stress caused by the fire would crack the windows and if the heat would degrade the window seals allowing radioactivity to escape from the cell and the oil reservoir which makes up part of the window would leak into the cell to add to the fire.

The heat transmitted to a window from burning heavy metal (HM) depends on the heat generated and the location of the material relative to the window. The window components and surrounding concrete are analyzed with a two-dimensional (depth and width of window) heat transfer model and a thermal stress model. The heat reaching the window from the fire is applied uniformly or with a distribution over the window surface. The distribution represents the thermal radiation view factor due to materials of different shapes.

The window can only receive heat from one side of the fire the maximum exposures 25% of the heat released. All this heat is applied directly to the window or the window and concrete (which does not account for heat loss to the cell gas by convection).

A thermal stress model is used to determine if the stress in any of the glass panes which make up the window exceeds the glass tensile stress strength which would cause cracking. The maximum amount of exposed HM (12 kg) was used as the heat source. The heat source was assumed to be comprised of dendrites that release a constant power until all the metal is reacted, which occurs in 31.8 minutes. The exposed metal may be in open canisters. A cylindrical geometry is assumed to result in a sine wave distribution of heat flux on the window. The cold side of a window pane will develop the tensile stress due to expansion of the hot side and therefore, is the side prone to cracking. The highest stress is on the cold side surface.

More concentrated HM in the form of a puck or button has much less surface area per unit volume than dendrites making the heat source density less than that of the dendrites (A denser heat source also has a higher ignition temperature and is less likely to burn.). A concentration of the heat flux was used to determine how much more stress HM burning in acylindrical container would provide than a uniform source. The seals are made of polyvinyl chloride (PVC) and can decompose (140°C) and melt at lower temperatures (160°C) compared to the other window components of concrete, metal, and glass. Thus, in some cases, the heat-flux term will be applied and concentrated near the window edge.

A previous study [1] developed for the facility safety analysis report [2] investigated the effect on the cell of exposing 62 kg of dendrites during a cell boundary breakage event. This amount generates 150 kW of power. The increased cell average wall temperatures (15°C increase) and cell atmosphere temperatures (35°C increase) were not enough to vaporize or decompose any material. There is no concern with a cell flashover to occur with the 150 kW heat release. For burning to occur (cell flashover), it is necessary to increase cell temperature enough to release gases from cell flammable material. The heat release from this incident 12 kg *150 kW/62 kg = 30 kW is less and would have less effect upon the cell than those considered in ECAR-534.

Radiant heat received on the glass surface on the interior of the cell is transmitted through the glass and concrete by conduction only. Although the glass window has a visible transmissivity of 38% to visible light, the majority of the heat produced will be in the infrared. In the infrared region (i.e., wavelengths greater than 0.8 micron), glass is opaque for wavelengths greater than 5 microns but transmits some radiation (perhaps 38%) for wave lengths from 0.8

to 5 microns. So some is transferred through the glass by radiation and some by conduction. This model assumes all the heat is passed by conduction. This assumption increases the temperature gradients in the glass and the thermal stress and is and so the model is very conservative. Heat loss back to the cell from the window is conservatively neglected. Combustion of the metal is assumed to be a constant rate until the metal is consumed. Reduction in the rate from oxide buildup could produce lower temperatures and stress.

2. METHOD

Equation (1) gives the conduction equation for the temperature T in two dimensions in region i as:

$$\rho_i C_{pi} \frac{\partial T_i}{\partial t} = k_i \frac{\partial^2 T_i}{\partial x^2} + k_i \frac{\partial^2 T_i}{\partial y^2} + q_i \tag{1}$$

where

ρ = density,

C_p = specific heat,

k = thermal conductivity,

q = heat generation rate per unit volume,

t = time,

x = horizontal axis,

y = vertical axis,

i = region or material.

The interface condition equating heat fluxes between two regions i and j is shown in Equation (2).

$$k_i \frac{\partial T_i}{\partial n} = k_j \frac{\partial T_j}{\partial n} \qquad (2)$$

where

n = x or y direction.

Equation (3) gives the boundary condition on the corridor side of the window is given as:

$$k_s \frac{\partial T}{\partial x} = h_c \left(T - T_c\right) \qquad (3)$$

where

k_s = thermal conductivity of Pane G,

h_c = heat transfer coefficient of the corridor air and a function of the forced circulation of the cooling system,

T_c = temperature of the corridor.

The temperature derivative is evaluated on the right-hand side of the boundary. The building air circulation system is assumed to provide adequate flow around the outside of the cell to yield a high-heat transfer coefficient estimated at 100 $W/m^2 \cdot K$. The value of 100 was chosen for a previous study inside the ZPPR vault which has a high wind velocity so the heat transfer coefficient is high. However, since the temperature increase never went further from the cell side than the B pane, the heat transfer coefficient on the corridor side is irrelevant and the same results would have been obtained with a zero or a value 10 times as high since no heat is transferred at that surface.

Equation (4) shows the two boundaries on the right and left of the window are treated as insulated as well the cell side of the concrete and steel not exposed to the heat flux.

$$\frac{\partial T}{\partial y}=0 \tag{4}$$

where y is the direction parallel to the cell wall.

In addition, the cell side of the window and concrete has a heat flux specified in Equation (5).

$$\frac{\partial T}{\partial n}=q \tag{5}$$

Equation (6) represents transverse thermal stress in a slab with a temperature distribution in the depth of the material. It is given in [3] and [4] as

$$\sigma_i=\frac{E\alpha}{1-\mu}\left(T_{avg}-T_i\right) \tag{6}$$

where

T_{avg} = average T in direction of the depth,

σ_i = stress at depth x_i,

E = Young's Modulus,

α = coefficient of thermal expansion,

μ = Poisson's ratio.

The thermal stress is calculated with a coefficient of thermal expansion in the glass of CTE = 9×10^{-6}/K, Young's modulus E = 6.89 × 10^4 MPa = 10 × 10^6 psi; and Poisson's ratio, μ = 0.16.Substituting the above values into Equation (6) gives the coefficient

$$\frac{E\alpha}{1-\mu}=115.3846\frac{\text{psi}}{K} \tag{7}$$

The tensile stress limit for glass is listed as 40 Mpa = 5714 psi [5] and [6] . Toughened glass limit is three to five times greater than the tension stress limit at 120 to 200 MPa. The minimum value of the tensile stress limit reported in the literature appears to be the 5714 psi so to be conservative this limit will be used to evalulate possible damage to the glass. The same limit is applied to all panes of glass in the window.

The compression limit is 1000 MPa = 142,900 psi. Compression of glass is usually not a problem and none of the cases investigated in this study exhibit significant compression stress.

Figure 1 shows the equipment which will be operated in the hot cell in front of the 11 and 12 windows. It is a demonstration kilogram-scale integrated pyrochemical process. It will provide an evaluation of the recycling of used LWR oxide fuels via electrochemical technologies and a fast reactor. This evaluation of electrochemical recycling will focus on the improvement of process knowledge such that the technology's benefits to an overall fuel cycle can be better quantified. It will provide material balance data, evaluate process flow sheet options, and serve as a test bed for demonstration of safeguards and security technologies. It also includes the fabrication of several metal fuel rodlets, their irradiation in the Advanced Test Reactor, and post-irradiation examination (PIE) in HFEF.

The cell perimeter shielding windows are layered glass and optical oil (mineral oil) and provide shielding approximately equal to that of the walls. Each window is composed of lead glass slabs with thin layers of oil between them, plus a protective glass plates on the cell side and the corridor side. The cell side protective plate is supported by a frame hinged from the cell liner so that it can be swung into the cell for cleaning or replacement with the remote handling equipment.

Type A-1 windows (from Schott Optical Glass, Co.) are located at the main and deconcell window positions 1M, 2M, 4M, 5M, 7M, 9M, 11M, 13M, 15M, and 3D (see Figure 2). Each consists of six slabs of glass, labeled A through F. A type A-1 window consists of a steel-tank assembly inserted from the cold side that is bolted to the out-of-cell side of the window-liner flange. The tank has a primary gas seal bolted to its hot face to mate with the liner flange. The tank assembly contains three heavy glass slabs, labeled C and D (cerium-stabilized), and E. A type A-1 window has a hinged protection plate (A) and is equipped with a secondary-seal assembly at the in-cell end. The hot end is sealed with a 9-in.-thick slab of cerium-stabilized glass (B) and the cold end is sealed with a 20-mm-thick plate of the same type of glass (F) but with no cerium (total glass weight is 6734 lb). The remaining free space is filled with approximately 55 gal of clear mineral oil.

Overall light transmission is 38% (including the hinged protection plate). Total thickness of the glass is 48 in. (the window affords 149 density-in. Density inches = [specific gravity] × [thickness in inches of shielding]). The clear aperture is 24-in.-high and 40-in.-wide.

Figure 1. Equipment layout for HFEF Windows 11M and 12M.

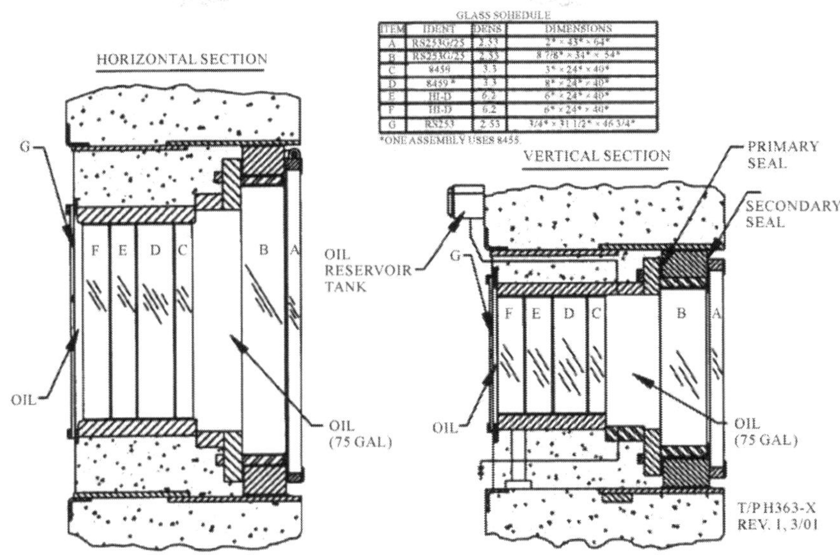

Figure 2. Type A-2 (Nuclear Pacific) radiation shielding window.

Drilled passageways in the liner flange are connected to a gas manifold that maintains an argongas pressure of 10-in.-wg. The primary and secondary seals (double seals with a center annulus) are made of extruded polyvinyl chloride. The center annulus is purged with argon gas in the primary seal and may be purged in the secondary seal if necessary. The center annulus is fed by the gas manifold.

The tank unit is supported near the cold end to prevent cantilevering the weight of the tank from the window-liner flange where the tank is bolted. During installation, a machine-leveling jack was placed under the support leg and adjusted to carry half the load. After installation and leak-checking the seals, the remaining annular space between the tank and window liner was carefully filled with a low-strength concrete grout with a specific gravity of 3.5 wg.

The oil reservoir (mounted on the operating corridor wall above the window) has a sight gauge with a capacity of about 1.6 gal of oil. Argon gas (at a pressure of 10-in.-wg) is piped into the free-air space in the top of the reservoir to prevent air from coming in contact with the window oil.

Type A-2 windows (from Nuclear Pacific, Inc.) are located at main and deconcell window positions 3M, 6M, 8M, 10M, 12M, 14M, 4D, and 5D. Each type A-2 window consists of seven slabs of glass, labeled A through G (see Figure 2). The Type A-2 window has been analyzed in this work but the results obtained apply to the Type A-1 window as well.

Viewing angles, shielding, primary and secondary seals, the oil reservoir, and grouting technique are the same for the type A-1 and A-2 windows. Tank structure is different (except for the flange at the hot end, which has to mate with existing window-liner flanges). The A-2 tank assembly contains four panes of glass (C, D, E, and F). The hot end of the tank is sealed with an 8-7/8-in.-thick slab of cerium-stabilized glass and the cold end is sealed with a 20-mm thickness of the same type of glass with no cerium. Total glass weight is 5933 lb. The remaining free space is filled with approximately 75 gal of clear mineral oil. Overall light transmission is approximately 38% (including the hinged protection plate, which is identical to that for a type A-1 window). The clear aperture is 24-in.-high and 40-in.-wide.A typical A-2 window is shown in Figure 2.

A typical window assembly is 64-in.-wide by 44-in.-high on the cell side, including the protective plate that tapers to 40-in.-wide by 24-in.-high on the corridor side.

Argon is contained between Panes A and B. Under cell pressure, no differential in pressure occurs from behind Pane A. Therefore, the cell pressure would not be applied to push the window out if a crack occurs.

75 gal of mineral oil is contained between PanB (8-7/8-in.-thick) and Pane C (3-in.-thick) with an estimated thickness of 11.625 in. A thin layer of oil is also between the remaining panes and maintained with an argon cover gas at a pressure of 10-in.-wg.There is concern that if the oil temperature increases then thermal expansion of the oil due to heating in the incident will cause an increase of pressure that cannot be relieved by the pressure system. This effect has been investigated in this study. The maximum oil temperature occurs in

scenarios 6 where the average increase is 0.03 C resulting in an increase of 0.0014 gallons of the original 75 gallons and is therefore not of concern. (Note, this temperature increase is obtained in the highest temperature case which is investigated later in the paper. This temperature rise is discussed later preceding Figure 23).

The dimensions of type A-2 window components are included in Table 1.

A horizontal cut through the centerline of the window is modeled in the analysis. A vertical cross-section model would produce similar results. The information input to the computer code is summarized in Figure 3. Note this figure is not to scale. For example, Panes C through F are all 24-in.-wide and only represented by a single box. The number of increments used in the solution is noted and each of those panes are subdivided into 8 increments (length) and 5 (width), which is 40 subareas. The material specifications are shown in Figure 4.

The material properties used corresponding these numbers are listed below.

m = 1: cp(m) = 0.24 * 4200: k(m) = 0.02 * 1: rho(m) = 0.001: A(m) = "Arg Nu=1" 'J/kg C, W/m K, gm/cc
m = 2: cp(m) = 880: k(m) = 1.1: rho(m) = 2.1: A(m) = "kconcrete" ' J/kg C, W/m K, gm/cc"
m = 3: cp(m) = 460: k(m) = 43: rho(m) = 7.85: A(m) = "ksteel"'J/kg C, W/m K, gm/cc"
m = 4: cp(m) = 1670: k(m) = 0.162: rho(m) = 0.8: A(m) = "oil" 'J/kg C, W/m K, gm/cc 24.200 J/mol K
m = 5: cp(m) = 840: k(m) = 1.05: rho(m) = 2.5: A(m) = "glas low dens"'J/kg C, W/m K, gm/cc"
m = 6: cp(m) = 840: k(m) = 1.05: rho(m) = 3.3: A(m) = "glas med dens"' J/kg C, W/m K, gm/cc"
m = 7: cp(m) = 840: k(m) = 1.05: rho(m) = 6.2: A(m) = "glas high densi" 'J/kg C, W/m K, gm/cc

Table 1. Dimensions of type A-2 glass window components.

Item	Identifier	Density	Dimensions
A	RS253G/25	2.53	2 in. × 43 in. × 64 in.
B	RS253G/25	2.53	8-7/8 in. × 34 in. × 54 in.
C	8459	3.3	3 in. × 24 in. × 40 in.
D	8459*	3.3	8 in. × 24 in. × 40 in.
E	HI-D	6.2	6 in. × 24 in. × 40 in.
F	HI-D	6.2	6 in. × 24 in. × 40 in.
G	RS253	2.53	3/4 in. × 31-1/2 in. × 46-3/4 in.

	In		Width (in)	0.75	0.50	6.00	0.25	6.00	0.25	8.00	0.25	3.00	11.625	8.88	0.50	2.00	Q
	cre		Increments	5	2	5	2	5	2	5	2	5	5	10	2	5	
	mer	Length		g	oil	f	oil	e	oil	d	oil	c	oil	b	arg	a	
1		(inches)		1.91	3.18	18.42	19.05	34.29	34.93	55.25	55.88	63.50	93.03	115.57	116.84	121.92	cm
6	5	12		Conc		Conc		Conc		Conc		Conc	Conc	Conc		Conc	30.48
8	2	5		Conc		Conc		Conc		Conc		Conc	Conc	St	St	St	43.18
10	2	4.5		Conc		Conc		Conc		Conc		Conc	St	St	Arg	A	54.61
12	2	1.25		St		Conc		Conc		Conc		Conc	St	B	Arg	A	57.79
14	2	3.75	Corridor	G	Oil	St	Oil	St	Oil	St	Oil	St	Oil	B	Arg	A	67.31
22	8	24	air	G	Oil	F	Oil	E	Oil	D	Oil	C	Oil	B	Arg	A	128.27
24	2	3.75	htcoeff	G	Oil	St	Oil	St	Oil	St	Oil	St	Oil	B	Arg	A	137.80
26	2	1.25		St		Conc		Conc		Conc		Conc	St	B	Arg	A	140.97
28	2	4.5		Conc		Conc		Conc		Conc		Conc	St	St	Arg	A	152.40
30	2	5		Conc		Conc		Conc		Conc		Conc	Conc	St	St	St	165.10
35	5	12		Conc		Conc		Conc		Conc		Conc	Conc	Conc		Conc	195.58
Sum	34		1	6	8	13	15	20	22	27	29	34	39	49	51	56	

Figure 3. Input geometry of the window assembly.

		1	2	3	4	5	6	7	8	9	10	11	12	13	
1 Argonne															
2 concrete	1	2	2	2	2	2	2	2	2	2	2	2	2	2	12
3 steel	2	2	2	2	2	2	2	2	2	2	2	3	3	3	5
4 Mineral c	3	2	2	2	2	2	2	2	2	2	3	3	1	5	4.5
5 glas 2.5	4	3	2	2	2	2	2	2	2	2	3	5	1	5	1.3
6 glas 3.3	5	5	4	3	4	3	4	3	4	3	4	5	1	5	3.8
7 glas 6.2	6	5	4	7	4	7	4	6	4	6	4	5	1	5	24
	7	5	4	3	4	3	4	3	4	3	4	5	1	5	3.8
	8	3	2	2	2	2	2	2	2	2	3	5	1	5	1.3
	9	2	2	2	2	2	2	2	2	2	3	3	1	5	4.5
	10	2	2	2	2	2	2	2	2	2	2	3	3	3	5
	11	2	2	2	2	2	2	2	2	2	2	2	2	2	12

Figure 4. Material specifications.

Properties of glass are given in [7] . Normal glass (referred to here as low density) specific heat and thermal conductivity were used for medium density and high density glass. As will be seen, since the heat from combustion did not have enough time to reach the medium and high density glass before combustion was completed, this assumption had no effect on the conclusions reached. Mineral oil properties are from [8] . Properties of PVC are taken from the Wikipedia web page [9] .

3. RESULTS

A summary of the physical situations and results is included in Table 2 with detailed results following.

1) Base case run with 12 kg dendrites burning (HFEFWindow1.xlsm).

The base case is a fire of 12 kg of uranium dendrites. The heat from one quarter or 3 kg is assumed to be applied directly to the entire Pane A. This amounts to 7.26 kW applied to the window. The area of the window is

Pane A = 43 *64 *2.54 *2.54 = 17,754.8 cm².

The heat flux per unit area is

$$\frac{Q}{A} = \frac{150\ \text{kW}}{62.18\ \text{kg}} * 3\ \text{kg} * \frac{1000\ \text{W}}{1\ \text{kW}} * \frac{1}{17754.8\ \text{cm}^2} * \frac{100\ \text{cm}}{\text{m}}$$

The temperature distribution in Pane A (the heavy lined area), the steel and concrete adjacent to it, the argon layer between Pane A and B, and the surface temperature of Pane B (heavy lined open box) is shown in Table 3. The top row is the distance from the corridor in cm, the next to the last column is the material in the last pane, and the last column is the surface heat flux incident on the window. All empty boxes in that column are considered insulated and after the completion of combustion, the window is also insulated. This temperature is at the maximum temperature just as the last of the metal uranium has reacted to oxide. The initial temperature of the window is specified to be 35°C. The temperature increase occurs in Pane A and very little in Pane B.

The temperature at the center of Pane A on the cell side of the glass versus time in minutes is shown in Figure 5. The above spatial distribution corresponds to the completion of combustion and maximum temperature. The temperature decreases sharply at the end of combustion but then drops slowly after combustion because the cell side of the glass is assumed insulated and all heat loss must eventually occur out of the corridor side of the window.

Table 2. List of cases investigated and results.

1. Base case run with 12 kg dendrites burning (HFEFWindow1.xlsm). The heat source is 3 kg * 150 kW/62.81 kg = 7.26 kW applied on the whole area. Pane A does not crack.

2. Base case with one half power for twice as long. (HFEFWindow2.xlsm). Decreased temperatures. Pane A does not crack.

3. Full power for twice as long (HFEFWindow3.xlsm). Increased temperatures in Pane A. Pane B small heat up. Stress not significantly increased in Pane A over base case. Stress in Pane B.

4. Concentrated heat flux cosine function heating (HFEFWindow4.xlsm). Center of heat flux centered on window. Stress is high enough that Pane A could crack. If Pane A stays in place, Pane B small heat up and small stress and will not crack.

5. Pane A disintegrated. Heat radiated directly to Pane B (HFEFWindow5.xlsm). Maximum was centered on center of window. Pane A and argon layer removed. Pane B could crack. If Pane A is open or removed, exposed material should be limited.

6. Maximum heat flux centered in location of the seal with Pane A removed (HFEFWindow6.xlsm). The seal would not degrade. Pane B could crack. If Pane A is open or removed, exposed material should be limited.

7. Maximum heat flux centered in location of the seal. Pane A in place (HFEFWindow7.xlsm). Pane A could crack. Pane B does not crack.

8. Pane A is assumed to break. Remainder of the heat is applied to Pane B (HFEFWindow8.xlsm). Pane A is assumed to break 1000 psi below the stress limit and is instantaneously removed. Remainder of heat is applied to Pane B. Pane B does not reach the stress limit and does not crack.

Table 3. Temperature distribution at the end of combustion.

115.57	116.205	116.84	117.856	118.872	119.888	120.904	121.92		
35.00003	35.00004	35.00004	35.00004	35.00005	35.00005	35.00005	35.00005		
35.00004	35.00004	35.00004	35.00004	35.00005	35.00005	35.00005	35.00005		
35.00066	35.00071	35.00075	35.00082	35.00088	35.00092	35.00094	35.00094	Conc	
35.01075	35.01149	35.01219	35.01332	35.01426	35.01494	35.01529	35.01529		
35.14047	35.14988	35.15903	35.17366	35.18599	35.19507	35.19995	35.19995		
36.3815	36.4721	36.5667	36.71808	36.84632	36.94585	37.00484	37.00484		
36.41457	36.50733	36.60421	36.75923	36.89057	36.9925	37.05293	37.05293	St	
35.97127	36.52058	36.63131	38.16231	38.54203	39.13358	39.94224	40.97364		40.35226
35.57231	62.16153	89.05606	89.8757	99.41035	118.0476	146.434	185.4796	glass A	40.35226
35.75516	65.9619	94.78178	95.6601	105.688	125.0912	154.162	193.2076		40.35226
37.83518	67.01755	96.37226	97.26688	107.4318	127.0477	156.3086	195.3543	glass A	40.35226
37.89629	67.27651	96.80122	97.70102	107.8886	127.5356	156.8197	195.8653		40.35226
38.07963	68.05339	98.0881	99.00344	109.2592	128.9991	158.3527	197.3984	glass A	40.35226
38.08543	68.08699	98.14468	99.06072	109.3185	129.0609	158.4164	197.462		40.35226
38.0947	68.14074	98.2352	99.15236	109.4134	129.1599	158.5181	197.5637		40.35226
38.09496	68.14258	98.23829	99.15549	109.4166	129.1632	158.5215	197.5671		40.35226
38.09497	68.14264	98.23838	99.15558	109.4167	129.1633	158.5216	197.5672		40.35226

38.09497	68.14264	98.23838	99.15558	109.4167	129.1633	158.5216	197.5672		40.35226
38.09496	68.14258	98.23829	99.15549	109.4166	129.1632	158.5215	197.5671		40.35226
38.0947	68.14074	98.2352	99.15236	109.4134	129.1599	158.5181	197.5637		40.35226
38.08543	68.08699	98.14468	99.06072	109.3185	129.0609	158.4164	197.462		40.35226
38.07963	68.05339	98.0881	99.00344	109.2592	128.9991	158.3527	197.3984	glass A	40.35226
37.89629	67.27651	96.80122	97.70102	107.8886	127.5356	156.8197	195.8653		40.35226
37.83518	67.01755	96.37226	97.26688	107.4318	127.0477	156.3086	195.3543	glass A	40.35226
35.75516	65.9619	94.78178	95.6601	105.688	125.0912	154.162	193.2076		40.35226
35.57231	62.16153	89.05606	89.8757	99.41035	118.0476	146.434	185.4796	glass A	40.35226
35.97127	36.52058	36.63131	38.16231	38.54203	39.13358	39.94224	40.97364		40.35226
36.41457	36.50733	36.60421	36.75923	36.89057	36.9925	37.05293	37.05293	St	
36.3815	36.4721	36.5667	36.71808	36.84632	36.94585	37.00484	37.00484		
35.14047	35.14988	35.15903	35.17366	35.18599	35.19507	35.19995	35.19995		
35.01075	35.01149	35.01219	35.01332	35.01426	35.01494	35.01529	35.01529	Conc	
35.00066	35.00071	35.00075	35.00082	35.00088	35.00092	35.00094	35.00094		
35.00004	35.00004	35.00004	35.00004	35.00005	35.00005	35.00005	35.00005		
35.00004	35.00004	35.00004	35.00004	35.00005	35.00005	35.00005	35.00005		

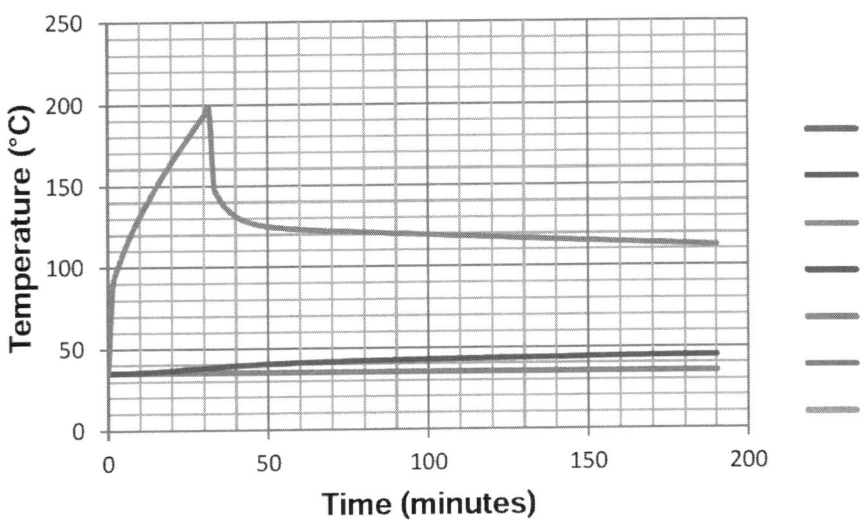

Figure 5. Temperature along the center of the glass.

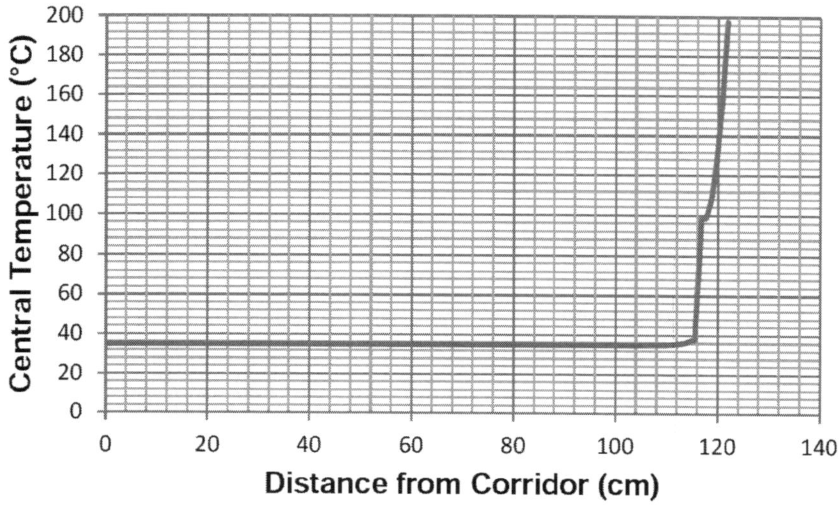

(Dimensions 93.03 cm; Pane B113.57 cm; argon116.84 cm; Pane A 121.92 cm)

Figure 6. Temperature distribution along the center of the window.

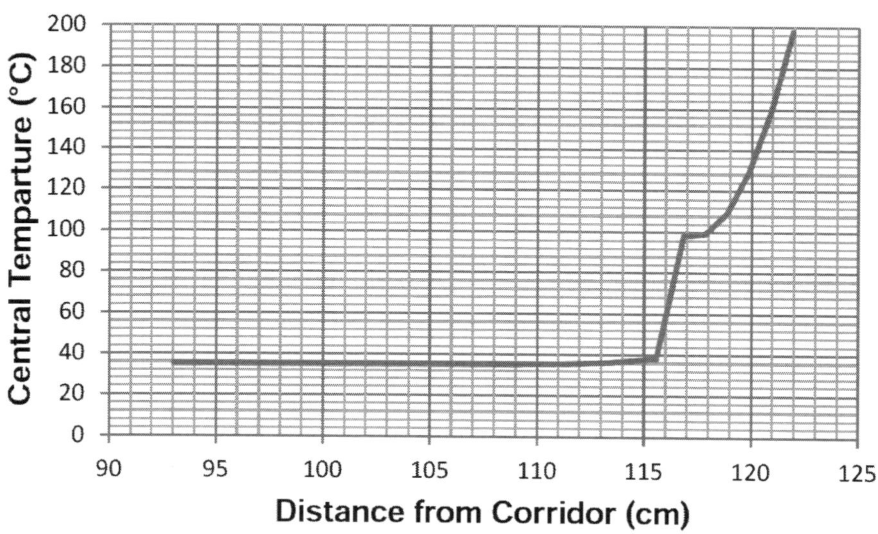

Figure 7. Temperature from corridor side of Pane B to cell side of Pane A.

Figure 6 emphasizes the fact that the temperature increase occurs in the argon between Pane A and B and Pane A.

Boundary locations between the materials are at the center of the window.

There is a small temperature rise in the cell side of Pane B.

An expanded view of the temperatures in Panes A and Bare shown in Figure 7.

The temperature distribution along the level of the window seal is shown in Figure 8. Just like the center temperature is almost flat in Pane B and starts to rise in the argon space, then rises in the Pane A glass. Note that the seal temperature is very low.

The calculated stress through the center of Pane A is shown in Figure 9.

As mentioned in the equation section, the tension stress limit for glass is listed as 40 MPa = 5714 psi. The calculated stress is less than the limit so cracking would not occur. The toughened glass limit is three to five times greater than the tension stress limit at 120 to 200 MPa. The compression limit is 1000 Mpa = 142,900 psi. Compression of glass is usually not a problem and is certainly not in this situation.

2) Base-case with one-half power for twice as long (HFEFWindow2.xlsm).

To investigate combustion at a lower rate, for example, due to oxygen starvation of the reaction in a can of dendrites, a run was made with half the power for twice the time for the same total heat flux. The maximum temperature is shown Figure 10. The maximum temperature is lower (160°C versus 200°C) and the stress is about half of the base case (2000 psi versus 4000 psi).

3) Full power for twice as long (HFEFWindow3.xlsm).

The base case was rerun assuming combustion continued for twice as long. The maximum temperature time history is shown in Figure 11. As expected, the temperature continues to rise until the end of combustion and reaches almost 290°C, which is 90°C higher than the base case.

The stress in Pane A is shown in Figure 12.

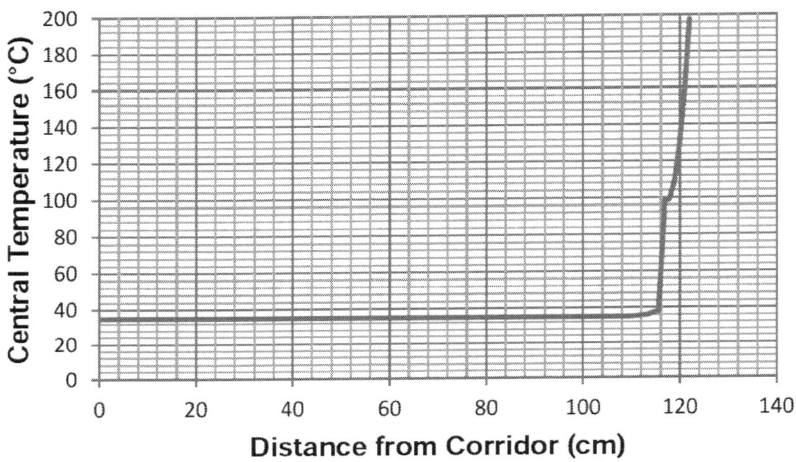

sealseal ← location of seal
Dimensions 93.03 cm Pane B 113.57 cm argon 116.84 cm Pane A 121.92 cm

Figure 8. Temperature along the level of the seal.

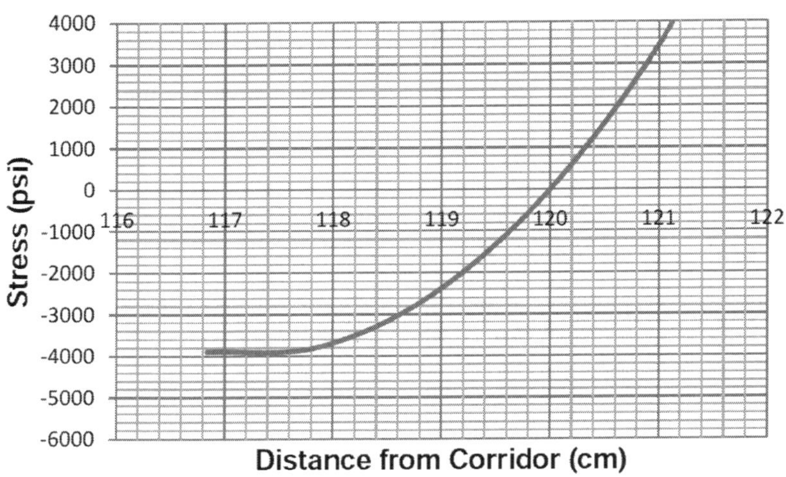

Figure 9. Stress levels in Pane A (Tension is negative and compression is positive.).

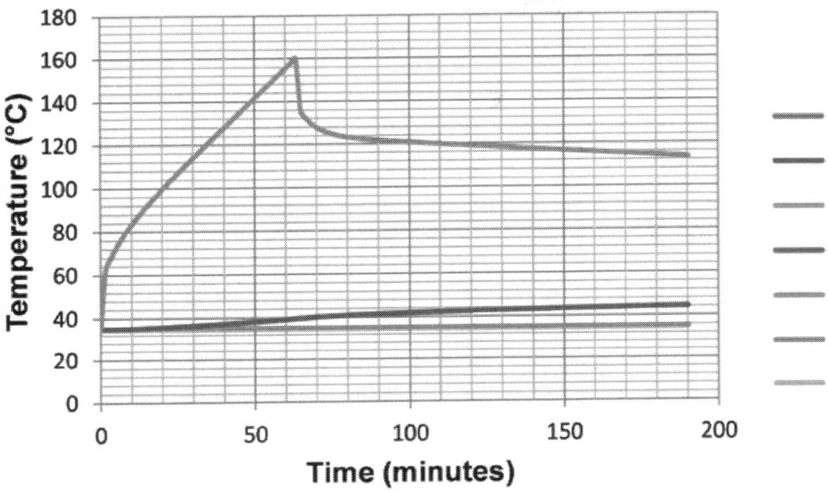

Figure 10. Investigation of effect of extending of the combustion time with the same total heat flux.

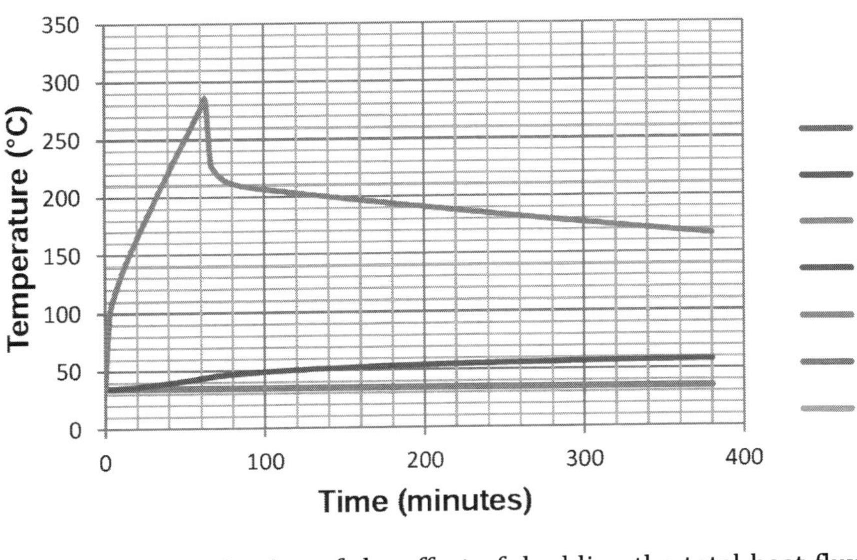

Figure 11. Investigation of the effect of doubling the total heat flux to the windows.

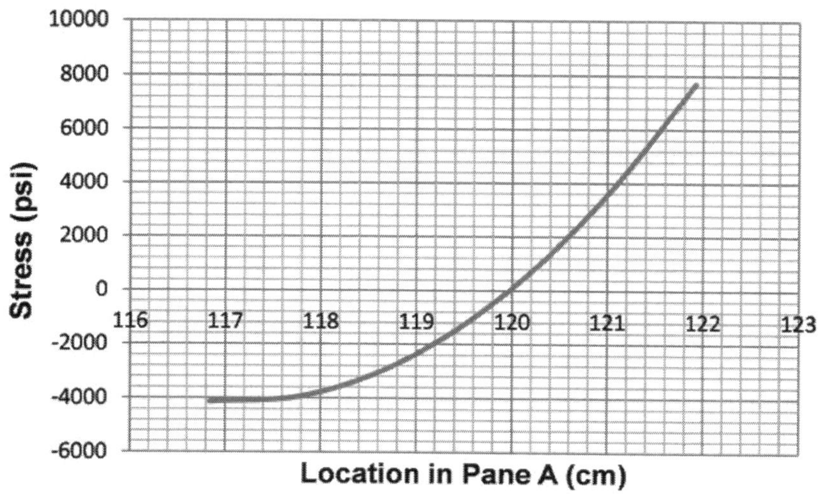

Figure 12. Stress in Pane A at the end of combustion.

The stress result is not much different than the shorter base case run. This is due to the whole temperature profile increasing in Pane A. That is, the shape is almost constant even though it is increasing and the stress depends on the temperature profile and not the absolute temperature. The temperature in Pane B heats up more than the base case, but the stress is still low (less than 200 psi).

These results imply that the maximum stress depends upon the rate of heat addition and not the total heat.

4) Concentrated heat flux cosine function heating (HFEFWindow4.xlsm).

The case with the same heat rate applied to the window but distributed with a cosine function was studied to determine the effect of the thermal radiation being concentrated as a result of view factors (e.g., if all the burning dendrites are in a canister).The same total heat flux was assumed to heat the window. The distribution shown in Figure 13 was assumed with the maximum being at the center of the window and the end from zero at the edge of the window.

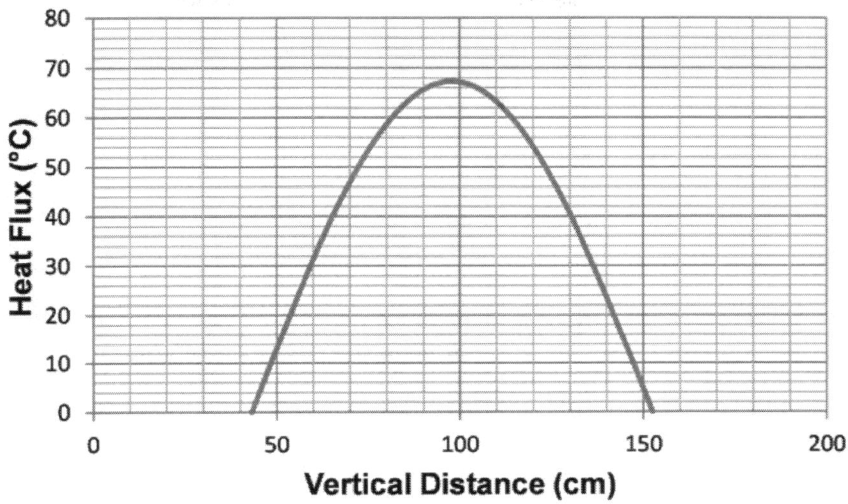

Figure 13. Cosine distribution of heat flux.

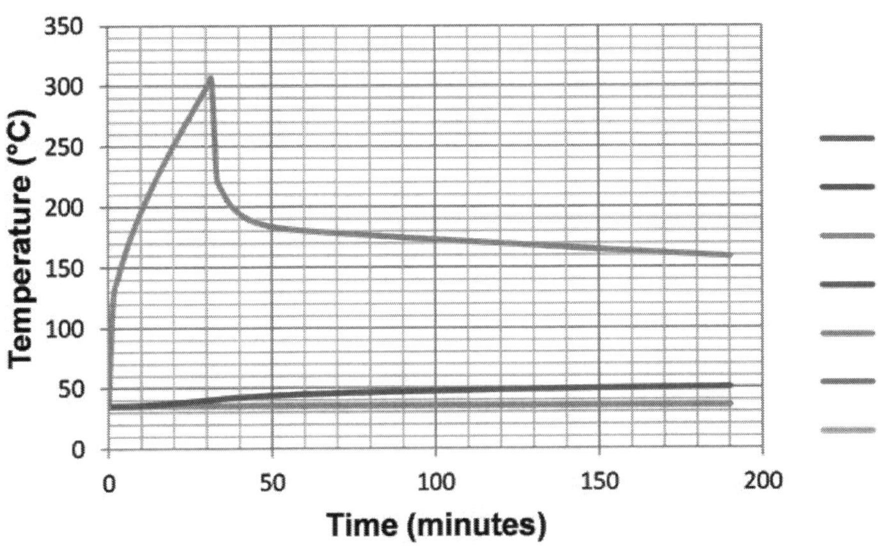

Figure 14. Maximum temperature with cosine heat flux distribution.

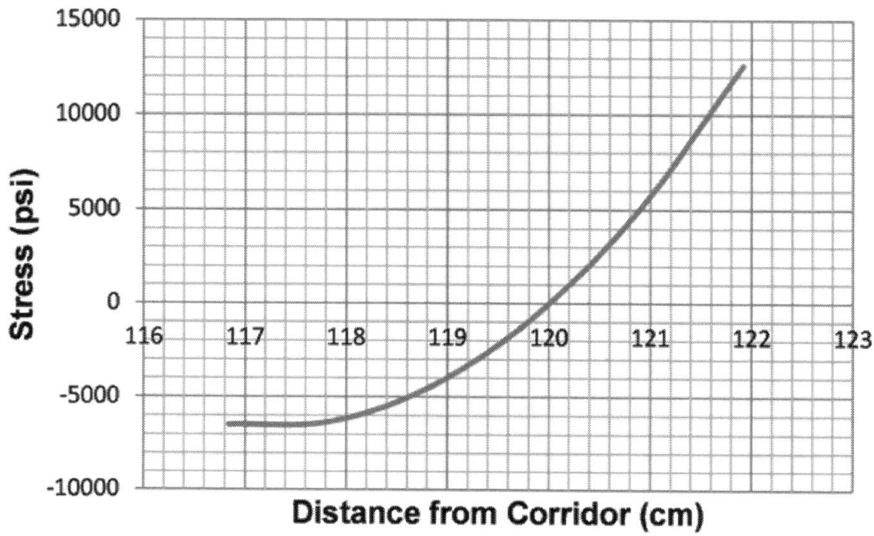

Figure 15. Stress distribution in Pane A for the cosine distribution.

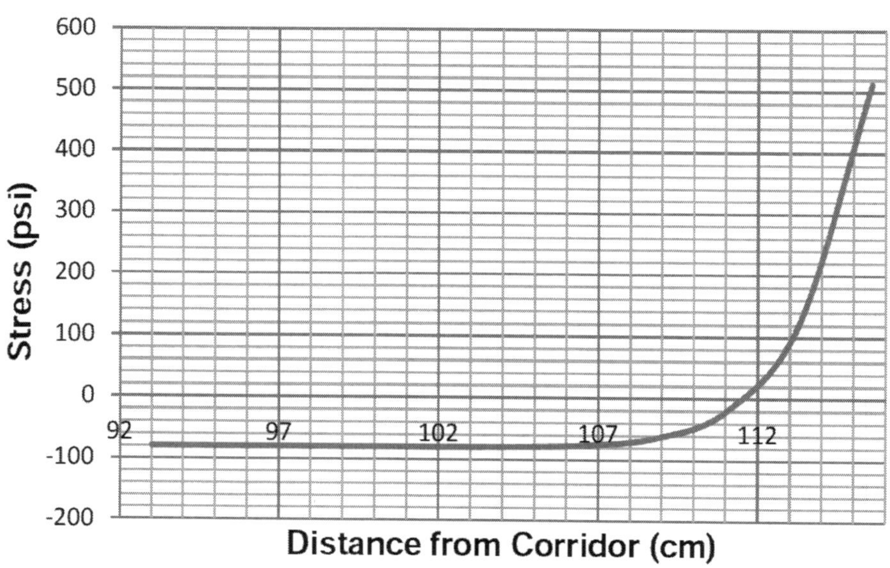

Figure 16. Stress in Pane B for the cosine distribution.

The temperature at the center of the window (shown in Figure 14) is higher by 100°C than the base case.

The stress along the centerline of Pane A is shown in Figure 15. The Pane Astress exceeds the tensile stress limit for a short time making it possible that this pane will crack.

This glass, even though cracked, could stay in place because of its thickness and the lack of a pressure difference between. The case where it completely falls out is investigated in a later calculation. If it stays in place, Pane B only heats up to 50°C. The stress in Pane B in this case is shown in Figure 16 and is seen to be far below the failure limit.

5) Layer A disintegrated allowing the heat to radiate directly to Pane B (HFEFWindow5.xlsm).

In order to assess the maximum possible danger to the HFEF window, the loss of Pane A was investigated. An analysis was conducted where Pane A was assumed disintegrated and removed. Heat radiated directly to Pane B. A cosine distribution of heat flux was centered on the window. The temperature distribution (shown in Figure 17) reached 340°C on the cell side of the Pane B.

Figure 18 shows the stress limit within the highest stress plane (through the center of Pane B) and exceeded the stress limit for almost 10 cm so it could crack.

The transient stress at the highest stress location (on the corridor side of Pane B) is shown in Figure 19. The stress exceeds the tensile stress limit for half the depth of Pane A only for a few minutes but it is possible that this pane may crack.

Item 8 analyzes the outcome of Pane A disintegrating and exposing Pane B during the transient. The results of this analysis indicate that the amount of exposed material should be reduced during operation if the Pane A is open or removed for programmatic reasons.

6) Maximum heat flux centeredon the seal with Pane A removed (HFEFWindow6.xlsm).

The cosine heat flux was applied with the maximum flux centered on the one side of the window seal with Pane A assumed to be open or removed. The transient temperature in the seal is shown in Figure 20.

Figure 21 shows the spatial profile at the seal horizontal location and the temperature on both sides of Pane B. The temperatures are low enough that no seal degradation is expected.

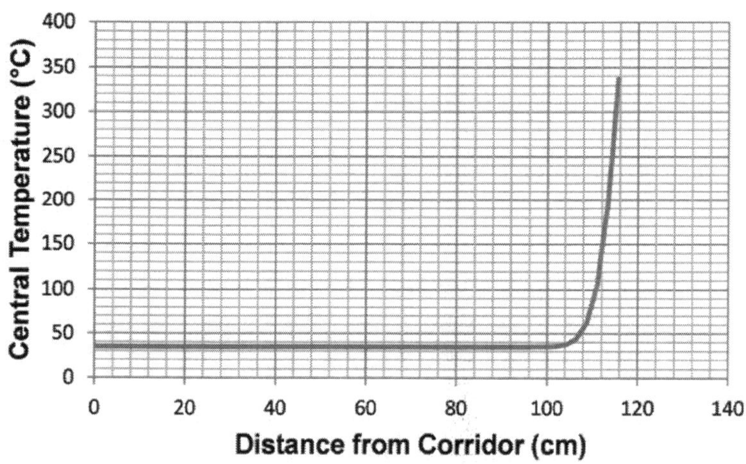

Figure 17. Temperature distribution with Pane A removed.

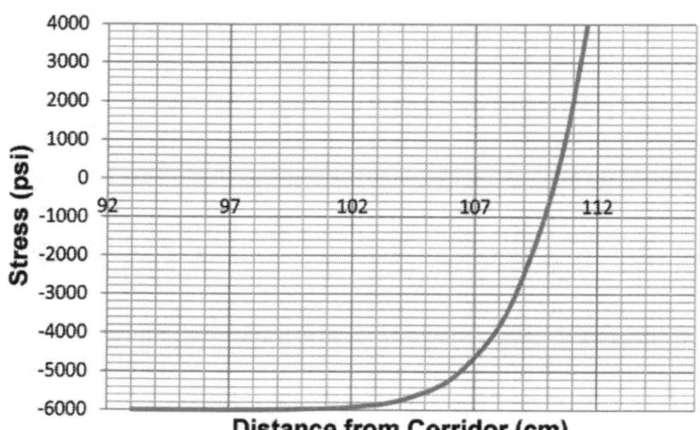

Figure 18. Stress in Pane B for the cosine distribution and with Pane A removed.

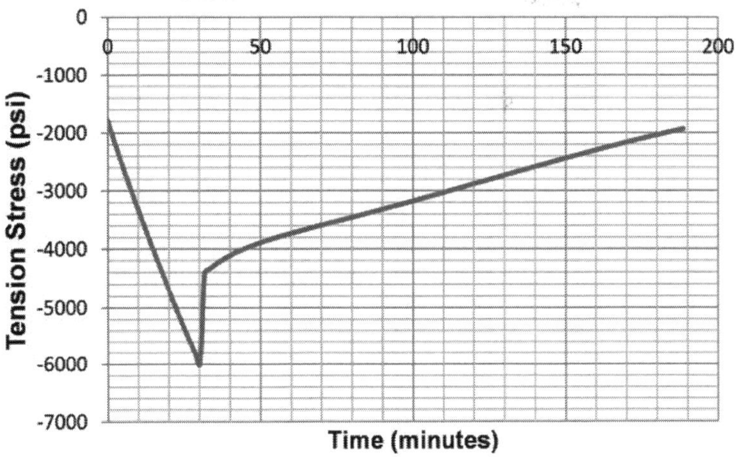

Figure 19. Transient stress on corridor side of Pane B.

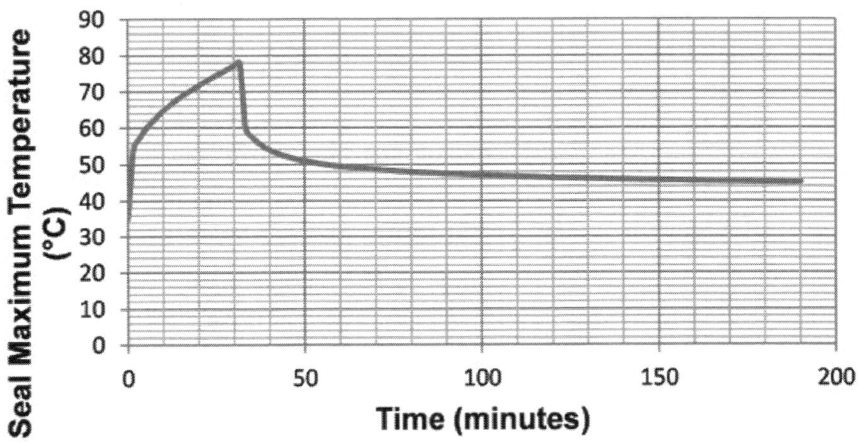

Figure 20. Transient seal temperature with Pane A removed.

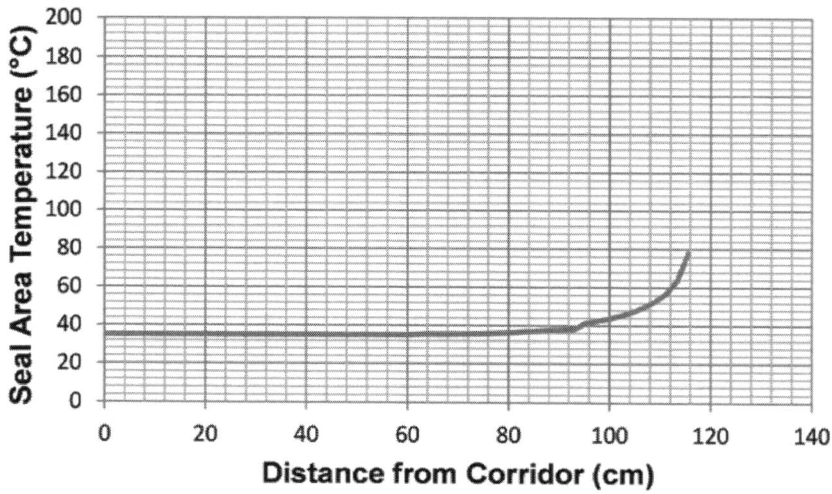

Figure 21. Temperature through seal area.

The stress through the seal area is shown in Figure 22. The stress (a few inches in) is a bit higher (<5200 psi) but lower throughout Pane B. The stress is not high enough to cause any damage to Pane B.

The oil temperature (Figure 23) increase is the highest in this scenario since the heat flux is centered near the edge of the glass. Heat can be conducted to the oil through the concrete better than though the glass. The maximum temperature occurs at the edge of the tank and is only 0.4 C above the starting temperature. The average temperature over the whole tank is only 35.03 C. The volumetric expansion coefficient [8] (Ref. Godfrey, 1995) of mineral oil is 6.4×10^{-4} C. The overall temperature increase of 0.03 C results in a volume increase of 0.0014 gallons over the original 75 gallons. Therefore the expansion of the oil is quite small and not of concern.

Also, the C pane (3 inches) is much thinner than the B pane (8.75 inches) so that it should crack before the B pane. This has in fact happened when the oil was pressurized with a positive displacement pump. In this case, any leaking oil would go toward the corridor instead of the cell.

7) Maximum heat flux centered on the seal with Pane A in place. (HFEFWindow7.xlsm)

The temperature at the seal horizontal location is shown in Figure 24. The seal temperature remains less than 50°C. No degradation of the seal is expected.

The stress profile in Pane A at the directly in front of the seal is shown in Figure 25. This is the maximum stress in Pane A. Pane A may crack because the stress levels exceed the stress limit. The stress in Pane B is low (<300 psi); there is no danger of it cracking.

8) Pane A is assumed to break. Then heat applied directly to Pane B (HFEFWindow8.xlsm).

The stress in Pane A is assumed to cause it to crack slightly below the tensile stress limit of 5714 psi. The transient stress at the center and on the corridor side of Pane A is shown in Figure 26. It is seen that the stress increases quite quickly but approaches a constant near the end of the transient. At 12.67 minutes and a stress of

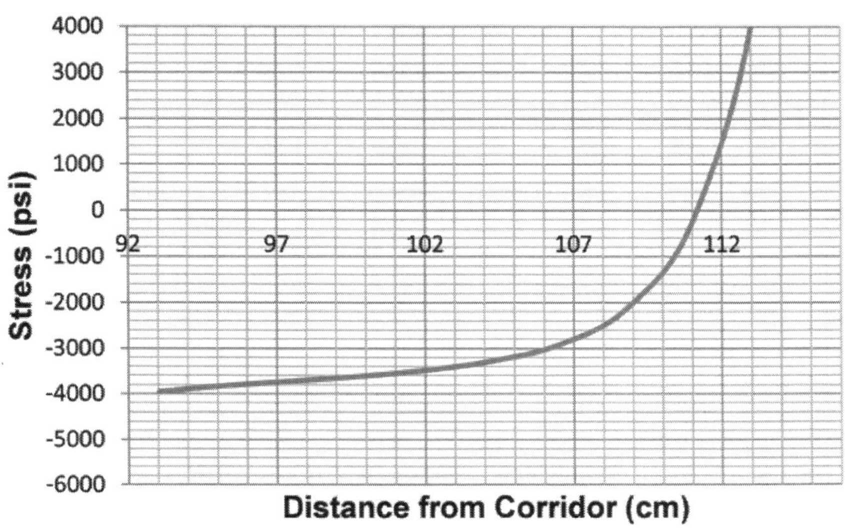

Figure 22. Stress throughout seal area.

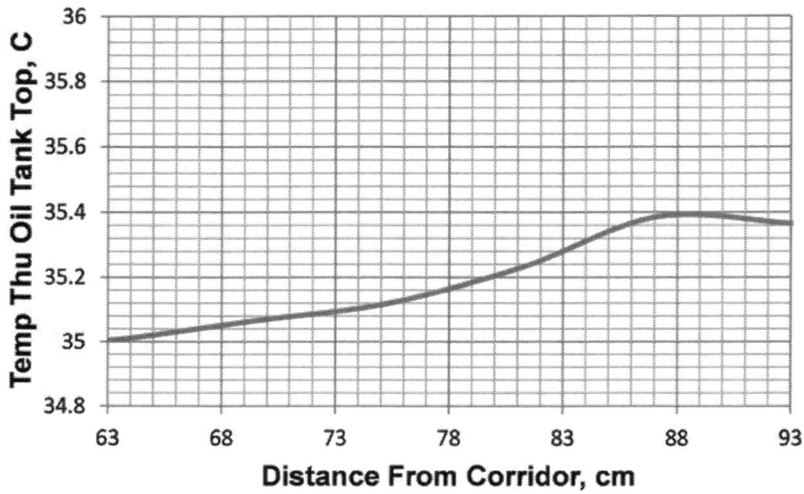

Figure 23. Oil temperature.

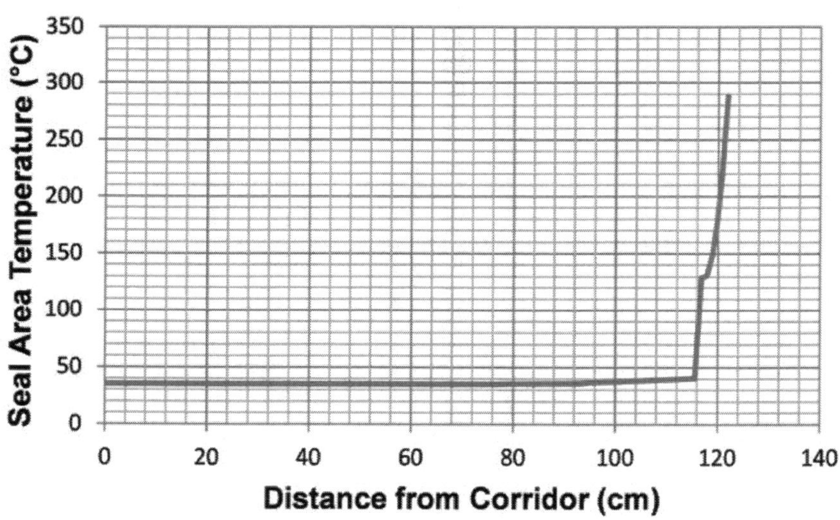

Figure 24. Temperature profile through the seal with Pane A (in place).

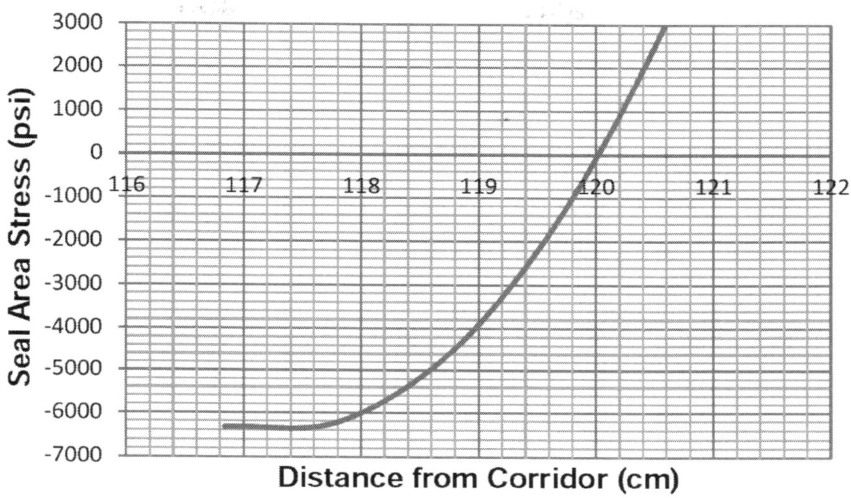

Figure 25. Stress in Pane A at seal area level.

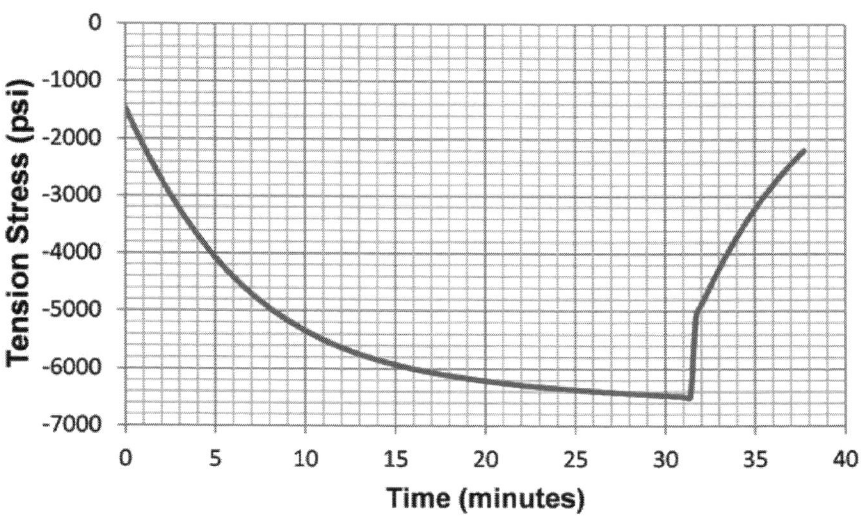

Figure 26. Stress in Pane A assumed to break at 5600 psi.

5600 psi, even though Pane A is 2 inches thick, it is assumed to rupture and completely disappear. Then all the heat from the fire is applied directly to Pane B.

Figure 27 shows the resulting transient stress at the center and on corridor side of Pane B.

The results in Figure 27 show the stress does not reach the tensile limit and only reaches 4400 psi.

As an additional conservatism, a case was run assuming that the tensile limit is actually 500 psi below the stated tensile limit. Pane A is assumed to fracture at 9.02 minutes at a stress of 5200 psi.

The stress in the center corridor side of Pane B when all the combustion heat is applied directly to Pane B at 9.02 minutes is shown in Figure 28. The stress is larger than the previous case but still remains below assumed lower stress limit of 5200 psi.

4. DISCUSSION OF RESULTS

The heat flux to the exposed window is quite high for the amount of metal assumed to be burning. It is assumed that all the material is dendrites (releases the highest heat flux of any material handled). It is more likely made-up of a mixture of lower-burning and high-burning rates. Also, it is unlikely that 12 kg will be exposed. One-fourth of the heat released is assumed to reach the window by thermal radiation. Due to the geometry, it is likely to be less. Also, convective losses to the cell environment are neglected. That is, if the metal fire itself were modeled in detail, the heat flux to the windows from 12 kg of burning dendrites would be less than assumed here. Also, the material will likely be distributed between two windows rather than concentrated in front of one window.

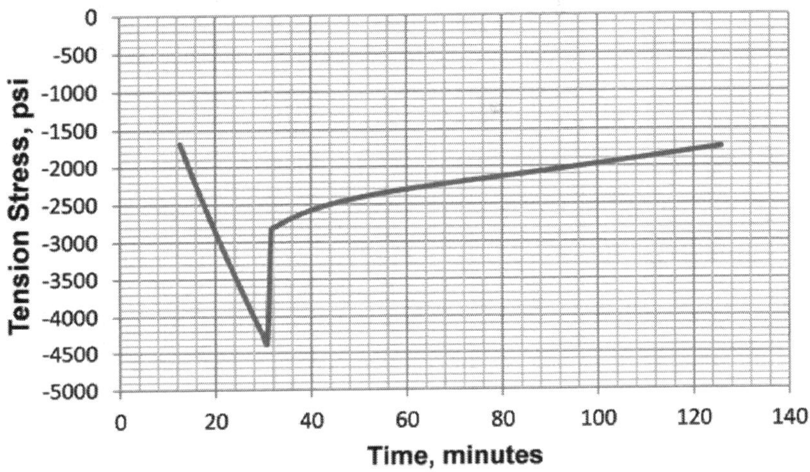

Figure 27. Stress at the mid plane on the corridor side of Pane B.

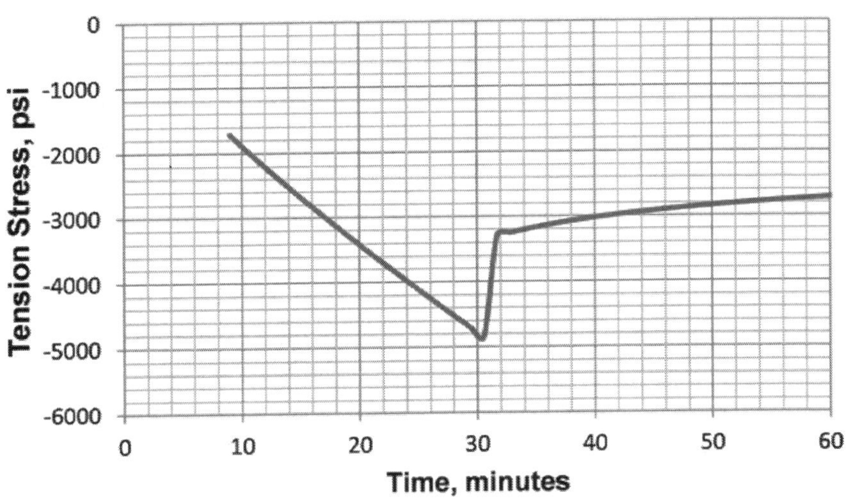

Figure 28. Midplane stress on the corridor side of Pane B assuming Pane A fractures at 5200 psi.

Most of the runs conducted were started with protective Pane A in place. When this pane is in place, the temperature rise barely reaches Pane B so there is no danger of Pane B breaking which is the window which forms a wall of the oil reservoir.

If Pane A completely disintegrates, it would expose Pane B to the heat flux at some point into the transient; however, since it is 2 inches thick, it is likely that it could remain in place even if cracked. The possible cracking of Pane B is significant because it is the barrier for the flammable mineral oil. None of the cases investigated indicate that Pane B will crack unless Pane A is removed or non-existent for the entire fire. If Pane A is missing, the amount of exposed material should be reduced.

In order to investigate the breaking of Pane A conservatively, it was assumed to be missing completely and the heat was applied directly to Pane B at the start of combustion. This also investigates the case if the fire happens if Pane A is open for cleaning. Pane B could crack when the heat is directly applied for the entire combustion event.

The heat flux was applied on the center of the seal. The seal did not heat up to 140°C where it could degrade. This is true whether or not Pane A is in place.PVC starts to decompose when the temperature reaches 140°C with melting temperature starting around 160°C.

A final case was investigated where Pane A was assumed to instantly disintegrate at the stress limit of 5700 psi and then the remaining heat was directly applied to Pane B. The thermal stress in Pane B did not exceed 4400 psi. A further case was run with Pane A assumed to disintegrate 500 psi below the stress limit at 5200 psi. The stress in Pane B remained below the stress limit reaching only 4900 psi.

5. CONCLUSIONS

The results indicate that in the worst case from a fire of 12 kg of dendrites during normal operation, Pane A could break and then expose Pane B to the remaining heat from the fire. Even in this case Pane B will not crack. This is the most stressful burning rate of the materials being processed. In the non-normal operation case where removable Pane A has been removed for repair or cleaning, Pane B would crack but would not break away and so probably would not allow much if any oil to leak from the window. In most operations, the amount of material exposed will be less than 12 kg and have lower heat-release rates so that Pane A will not break.

Since Pane B will not break, mineral oil will not be released into the cell so it will not ignite and add heat to the cell A previous study showed that combustion of a much larger amount of dendrites, 62.81 kg, only raises the cell atmosphere temperature by 35°C. This temperature level is not high enough to cause the release of volatiles from other materials; therefore, no flashover of fire in the cell is possible from the smaller 12 kg heat source considered in this study.

The analysis presented here assumed that all the heat reaching the window was transmitted in the window by conduction but in reality, some would be transmitted through the glass to the surrounding corridor. The assumption of no transmission through the window increases the temperature gradients in the glass panes which increases the thermal stress. Thus the cases which showed pane A cracking and breaking away, in reality probably would not occur because the heat reducing effect of the transmission of radiant energy through the window was not taken into account.

ACKNOWLEDGEMENTS

This work is supported by the US Department of Energy, Office of Nuclear Energy, under DOE Idaho Operations Office Contract DE-AC07-05ID14517. Accordingly, the US Government retains a nonexclusive, royalty-free license to publish or reproduce the published form of this contribution, or allow others to do so, for US Government purposes.

DISCLAIMER

expressed or implied, or assumes any legal liability or responsibility for the accuracy, completeness, or usefulness of any information, apparatus product, or process disclosed, or represents that is use would not infringe privately owned rights. References herein to any specific commercial product, process, or service by trade name, trademark, manufacturer, or otherwise, do not necessarily constitute or imply its endorsement, recommendation, or favoring by the US Government or any agency thereof. The views and opinions of authors expressed herein do not necessarily state or reflect those of the US Government or any agency thereof.

CONFLICT OF INTERESTS

The authors declare that there is no conflict of interests regarding the publication of this paper.

REFERENCES

1. Solbrig, C.W. (2009) Mathematical Model of the Hot Fuel Examination Facility Main Cell Response to Breaches. ECAR-534, Idaho National Laboratory.
2. SAR 405, Safety Analysis Report for the Hot Fuel Examination Facility, Enclosure 2, 2014.
3. Timoshenko, S. and Goodier, J.N. (1970) Theory of Elasticity. 3rd Edition, McGraw-Hill, New York.
4. Solbrig, C.W. and Bateman, K.J. (2010) Modeling Solidification-Induced Stress in Ceramic Waste Forms Containing Nuclear Wastes. Nuclear Technology, 172, 189-203.
5. (Gobain) Saint Gobain Glass UK, Glass Mechanical Properties. http://uk.saint-gobain-glass.com/trade-customers/physical-properties
6. (Lehman) The Mechanical Properties of Glass, Glass Engineering 150:312 Professor Richard Lehman, Department of Ceramics and Materials Engineering, Rutgers University,

New Brunswick. http://glassproperties.com/references/MechPropHandouts.pdf

7. Shand, E.B. (1958) Glass Engineering Handbook. 2nd Edition, McGraw-Hill, New York.
8. Godfrey, D. and Herguth, W.R. (1955) Physical and Chemical Properties of Industrial Mineral Oils Affecting Lubrication, Part 2. Journal of the Society of Tribologists and Lubrication Engineers.
9. (PVC Web Page) Polyvinyl Chloride Thermal, Melting and Decomposition Properties Taken from Web Page. http://en.wikipedia.org/wiki/Polyvinyl_chloride#Physical_properties

6

CHAPTER

A NUMERICAL STUDY ON SYSTEM PERFORMANCE OF GROUNDWATER HEAT PUMPS

Jinsang Kim [1] and Yujin Nam [2,*]

[1] *Blue Economy Strategy Institute, Suite 911, 55 Digital-Ro 34-Gil, Seoul 08378, Korea*

[2] *Department of Architectural Engineering, Pusan National University, Jangjun-2 Dong, Busan 609-735, Korea*

ABSTRACT

Groundwater heat pumps have energy saving potential where the groundwater resources are sufficient. System Coefficients of Performance (COPs) are measurements of performance of groundwater heat pump systems. In this study, the head and power of submersible pumps, heat pump units, piping, and heat exchangers are expressed as polynomial equations, and these equations are solved numerically to determine the system performance. Regression analysis is used to find the coefficients of the polynomial equations from a catalog of performance data. The cooling and heating capacities of water-to-water heat pumps are determined using Energy Plus. Results show that system performance drops as the water level drops, and the lowest flow rates generally achieve the highest system performance. The system COPs are used to compare the system performance of various system configurations. The groundwater pumping level and temperature provide the greatest effects on the system performance of groundwater heat pumps along with the submersible pumps and heat exchangers. The effects of groundwater pumping levels, groundwater temperatures, and the heat transfer coefficient in heat exchanger on the system performance are given and compared. This analysis needs to be included in the design process of groundwater heat pump systems, possibly with analysis tools that include a wide range of performance data.

Keywords: groundwater heat pump; coefficient of performance; regression analysis; UA value; heat exchanger

1. INTRODUCTION

A large portion of energy use in buildings goes towards heating and cooling. Operating buildings contributes more to energy use and climate change than either transportation or industry, according to an estimate of global energy consumption [1]. On the other hand, buildings offer the largest global potential to reduce greenhouse gas consumption [2]. Geothermal heat pumps are the most environmentally friendly and efficient method for heating and cooling buildings, and thus could have the largest potential to mitigate greenhouse gas emissions.

Generally, ground heat source systems utilize the annually stable underground temperature. In particular, Groundwater Heat Pumps (GWHPs) directly use groundwater that has a huge amount of potential as a heat source. The Equitable Building—located in Portland (OR, USA)—initiated the use of groundwater as a heat source for heating and cooling in 1948; in recognition, the American Society of Mechanical Engineers recorded it as a National Mechanical

Engineering Landmark. It is possibly the oldest geothermal heat pump and the oldest commercial heat pump.

In Korea, the use of renewable energy systems, including ground source heat pump systems, has been spreading widely in the building sector since it became mandatory in public buildings in 2004. Figure 1 shows applications of ground source heat pump systems in Korea [3,4]. According to a market analysis [5], the market for ground source heat pump systems in Korea may grow to 500 billion dollars. Although the market has been growing dramatically, most systems were closed loop system using a U tube pipe as a ground heat exchanger. GWHP systems are used in large-scale buildings, but it is difficult to say that the systems have spread widely.

There are several barriers to the installation of a GWHP system, such as the limitations of groundwater condition, pumping power, and local regulations. In particular, most building designers did not recognize the system as a heating and cooling system, and did not understand the proper design methods. Although the Korean government has suggested design and installation guidelines for ground source heat pump systems [6], the explanations for GWHP systems are not sufficient.

Proper design of GWHP systems can realize both low installation and operation costs [7]. While ground-source heat pumps can be installed almost anywhere, GWHPs can be successfully constructed and operated only where an aquifer can sustainably provide groundwater to the heat pump units. Adequate site characterization and system design are essential for stable operation and efficiency. Several studies have been conducted on the design factors of GWHP systems. Nam and Ooka conducted a numerical simulation of ground heat and water transfer for GWHP systems compared with real-scale experiments [8]. The results confirmed that the groundwater flow and the well position are important for the optimum design of GWHP systems, and the electricity consumption of a pump must be reduced significantly to achieve a higher system coefficient of performance

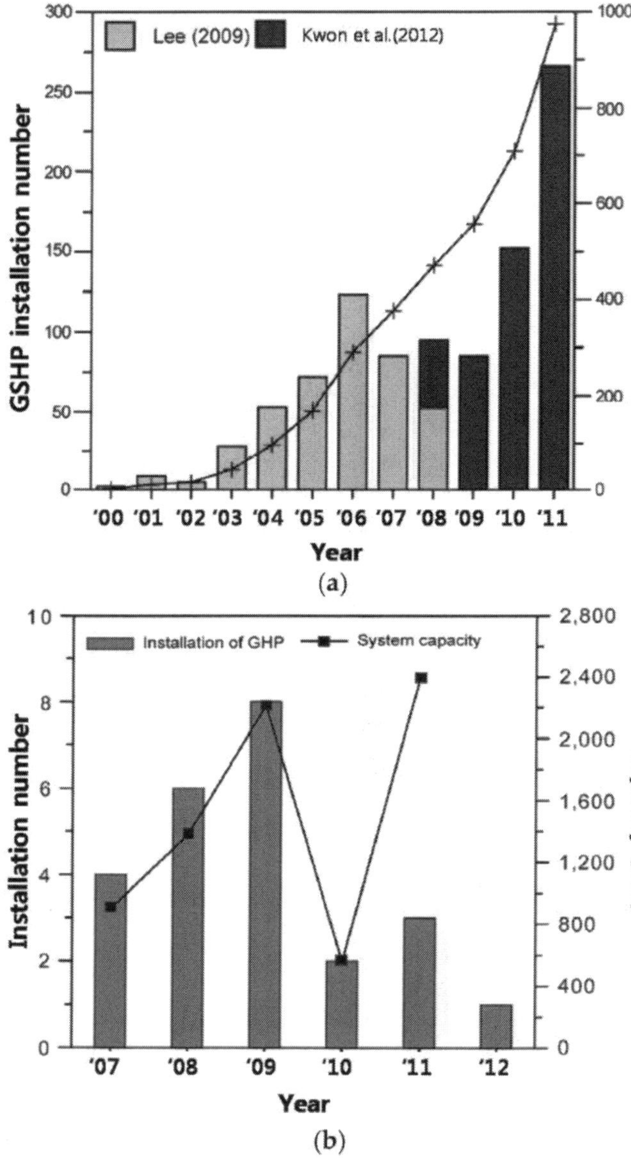

Figure 1. Current status of ground source heat pump system use in Korea. (**a**) Application of ground source heat pump systems; (**b**) Application of groundwater heat pump systems [3,4].

(COP). Zhou et al. also conducted numerical and experimental investigations of the thermal behavior in the well of a GWHP [9]. They compared the changes in the pumping temperature in different aquifer advection configurations and found that horizontal downstream cross-flow can improve the system efficiency in heating and cooling. ASHRAE (American Society of Heating, Refrigerating and Air-Conditioning Engineers) Research Report 1119 focused on the standing column well (SCW) design, and a design method for SCW systems was developed [10]. In that research, a simplified simulation tool for SCW systems was developed, and a parametric study for the design guidelines was conducted. Through the simulation tool, the groundwater temperature can be calculated for various conditions of the specific system and operation. Although the ground temperature or circulation temperature is important for calculating the performance of a heat pump, from the perspective of total system performance, the design and selection of the pumping system is also important.

Pumping systems and their operating conditions greatly influence GWHP system performance. Groundwater usually provides favorable operating conditions for heat pump units as compared to ground-source heat pumps, with higher COPs for heat pump units in GWHP systems. However, open-loop groundwater pumping power consumption increases rapidly as the groundwater level drops.

This theory is very simple and well known, but, in real system applications, it is difficult to quantitatively consider the relationship between the system performance and the groundwater conditions. Most system designers did not consider the COP of the total system, but only the lift of the pump for groundwater level and the friction factor. Of course, numerical simulations with ground heat and groundwater transfer models can support optimum designs, but they also have the limitation of practical calculation time. In order to develop a simple and practical design guide, it is necessary to utilize available performance data and individual component models.

In this study, the efficiencies of GWHP systems are analyzed based on the system heating and cooling COPs. The system COPs are calculated by employing polynomial models for heat pump units and pumps. The coefficients of the polynomial models are found by applying regression analysis to the manufacturers' catalog data. The effects of groundwater level and temperature, as well as heat exchanger UA (overall heat transfer coefficient) value (i.e., the rate of heat transfer per temperature difference), on the system performance are analyzed. The analysis will be helpful in determining system performance during the design process of GWHP systems, possibly with analysis tools that include a wide range of heat pump units and pump performance data.

2. GROUNDWATER HEAT PUMP (GWHP) SYSTEM

GWHP systems may consume a smaller amount of energy in operating heat pump units than ground source heat pump (GSHP) systems because their heat pump units usually operate under more favorable operating conditions, i.e., groundwater temperatures that remain stable throughout the year. However, in some instances, GWHP systems could consume more energy for pumping groundwater. The pumping energy consumption increases greatly as the groundwater level drops or the pumping flow rate increases.

The water table is maintained in wells and unconfined aquifers when no water is pumped from them. The static water level is the distance from the ground surface to the water level in the well. When water is pumped from a well, the water level in the well drops from the static water level. The water level in the well while pumping is called the pumping water level. The distance between the static water level and the pumping water level is called the drawdown [11,12]. The pumping water level is used in calculating the GWHP system performance.

GWHP systems are composed of one or more production water wells, heat pump units, submersible and circulation pumps, and heat exchangers. Their layouts can vary significantly with design intents and the number of production wells, heat exchangers, and other components. Some GWHP systems do not include heat exchangers between the heat pump units and the submersible pumps, allowing the groundwater to pass through the heat pump unit. This configuration is called a direct GWHP system, and is sometimes used in smaller capacity applications. Indirect GWHP systems, which include a heat exchanger between the groundwater submersible pump and the heat pump units, are often used in large commercial systems.

In this study, the indirect GWHP system is composed of one groundwater production well, one submersible pump, one water-to-water heat pump unit, one circulation pump, and one heat exchanger, as shown in Figure 2. Air handling units (AHUs) or fan coil units (FCUs) that are connected to the water-to-water heat pump unit are not included in calculating the system performance of the current GWHP system. The flow rates and inlet temperature to the AHUs are predefined and assumed to be unchanged. The system configuration is the same as that from the authors' preceding publication [13].

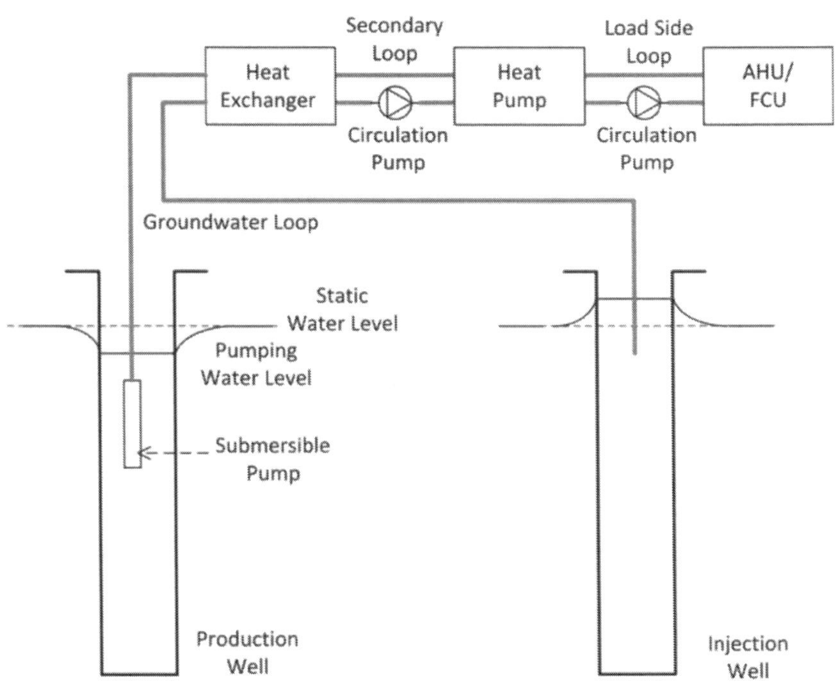

Figure 2. Groundwater heat pump system layout.

The submersible pump installed in the production well pumps the water, transfers it through the heat exchanger, and generally sends it to the injection well (unless there is another application). When the heat pump unit is in heating mode, the heat of the groundwater is transferred to the circulation water in the heat exchanger. The circulation fluid from the heat exchanger is sent to the heat pump unit by a circulation pump in the secondary loop. The heat of the circulation water is transferred to the refrigerant loop inside of the heat pump unit. The circulation water from the heat pump then returns to the heat exchanger in the secondary loop. The heat pump unit produces hot water with the capacity including the absorbed heat from the circulation water in the secondary loop and the electricity consumed by the compressor. The hot water produced by the heat pump units is sent to the AHUs or FCUs to heat the space, or for another process. In heating mode, the heat from the groundwater moves to the space or the process through the heat exchanger, heat pump unit, and the AHUs.

In cooling mode, the heat of the space or the process is transferred to the refrigerant loop inside of the heat pump unit via the AHUs and the circulation water in the load side loop. The circulation water in the secondary loop is chilled by the heat pump unit. Inside the heat exchanger, the heat travels from the circulation water to the groundwater, and the circulation water thus becomes cooler and the groundwater becomes hotter. The groundwater from the injection well is heated in the heat exchanger, and returns to the injection well (unless it is sent to another application).

3. COMPONENT MODELS

The GWHP system of this study is composed of one submersible pump installed in the groundwater production well, one water to water heat pump unit, one circulation pump, and one heat exchanger. In order to calculate the power consumption and the heating and cooling capacities of a GWHP system, the power consumption, head, and heating and cooling capacities of the components need to be modeled as functions of flow rate and temperature. When the flow rate varies in a GWHP system, the temperature distributions in the heat exchanger and heat pump unit vary. Accordingly, the heating and cooling capacities of the GWHP have different values.

3.1. Submersible Pump and Circulation Pump Model

Submersible pumps generally consist of multiple stages of identical impellers and housings installed in series with electric motors installed at the ends. The head and power consumption of submersible pumps are calculated by multiplying the number of stages n of the pump and the associated values for a single stage:

$$H_p = n \cdot h_p \tag{1}$$

$$W_p = n \cdot w_p \tag{2}$$

The head and power consumption for a single stage of a submersible pump are modeled in cubic polynomial equations in terms of flow rate V (in LPM (liters per minute)):

$$w_p = A_0 + A_1 V + A_2 V^2 + A_3 V^3 \tag{3}$$

$$W_p = B_0 + B_1 V + B_2 V^2 + B_3 V^3 \tag{4}$$

The eight coefficients in Equations (3) and (4) can be obtained through regression analysis from performance data typically provided in the manufacturer's catalog. For the single stage, h_p and w_p are the head (in m) and power consumption (in kW), respectively. The groundwater production well is assumed to be capable of providing the necessary flow rate to the heat exchanger or the heat pump unit, and it maintains the pumping water level throughout the operation. The circulation pumps in the secondary loops have similar characteristics as the submersible pumps. The same equation forms for the head and power consumption are used.

3.2. Heat Pump Model

When the circulation fluid of the secondary loop passes through the heat exchanger installed inside the heat pump unit to exchange heat with the refrigerant, head loss and temperature change occurs in the circulation fluid. The head loss in the secondary circulation loop of the heat pump unit H_{hp} is expressed in the following equation. The same equation is used for the head loss that occurs in the load side loop of the heat pump unit:

$$H_{hp} = C_{hp} \cdot V^2 \tag{5}$$

Temperature variations inside the heat exchanger and flow rate variations in the secondary loop cause the heat pump unit to produce different cooling and heating capacities—Q_c and Q_f (in kW), respectively—and consume different amounts of electricity—W_c and W_f (in kW), respectively:

$$\frac{Q_c}{Q_{c,ref}} = D_0 + D_1\left(\frac{T_{S,in}}{T_{ref}}\right) + D_2\left(\frac{T_{L,in}}{T_{ref}}\right) + D_3\left(\frac{V_S}{V_{ref}}\right) + D_4\left(\frac{V_L}{V_{ref}}\right) \tag{6}$$

$$\frac{Q_h}{Q_{c,ref}} = E_0 + E_1\left(\frac{T_{S,in}}{T_{ref}}\right) + E_2\left(\frac{T_{L,in}}{T_{ref}}\right) + E_3\left(\frac{V_S}{V_{ref}}\right) + E_4\left(\frac{V_L}{V_{ref}}\right) \tag{7}$$

$$\frac{W_c}{W_{c,ref}} = F_0 + F_1\left(\frac{T_{S,in}}{T_{ref}}\right) + F_2\left(\frac{T_{L,in}}{T_{ref}}\right) + F_3\left(\frac{V_S}{V_{ref}}\right) + F_4\left(\frac{V_L}{V_{ref}}\right) \tag{8}$$

$$\frac{W_h}{W_{h,ref}} = G_0 + G_1\left(\frac{T_{S,in}}{T_{ref}}\right) + G_2\left(\frac{T_{L,in}}{T_{ref}}\right) + G_3\left(\frac{V_S}{V_{ref}}\right) + G_4\left(\frac{V_L}{V_{ref}}\right) \tag{9}$$

Subscripts S and L indicate the secondary loop and the load side loop, and subscript ref is the reference value for the corresponding variable. The performance data of the water-to-water heat pump unit can be used to find the coefficients through regression analysis. The performance table includes the cooling and heating capacities and power consumption for source and load side water temperatures and flow rates [14].

3.3. Piping Model

Head loss occurs when the fluid passes inside the pipes and piping accessories. Pipe diameters, pipe lengths, and fluid properties influence the heat loss that occurs in the secondary and groundwater loops. The equation of the head losses expressed in terms of flow rate is as follows:

$$H_{piping} = C_{piping} \cdot V^2 \tag{10}$$

3.4. Heat Exchanger Model

Plate frame plate heat exchangers are normally used as heat exchangers installed between heat pump units and groundwater submersible pumps in indirect GWHP systems. The heat transfer occurs between the groundwater and the circulation fluid of a secondary loop inside the heat exchanger. When the flow rate of the groundwater or circulation fluid changes, the output temperatures vary in the following way:

$$H_{hx} = C_{hx} \cdot V^2 \tag{11}$$

$$Q_{hx} = \rho c_p \cdot \Delta T \cdot V \tag{12}$$

$$Q_{hx} = \mathrm{UA} \cdot \Delta T_{LM} \tag{13}$$

$$\Delta T_{LM} = \frac{\Delta T_1 - \Delta T_2}{\ln\left(\frac{\Delta T_1}{\Delta T_2}\right)} \tag{14}$$

Subscript hx indicates the heat exchanger, UA is a heat exchanger characteristic, and ΔT_{LM} is the logarithmic temperature difference of the heat exchanger as defined in Equations (13) and (14). U value is overall heat transfer coefficient, which considers the heat transfer coefficients of circulation fluid and the resistance of the pipe material.

4. GWHP SYSTEM CONFIGURATION

The indirect GWHP system layout in this study is composed of one submersible pump installed inside the groundwater production well, one water to water heat pump unit, one circulation pump, and one heat exchanger.

A 4 inch diameter submersible pump SP 75S made by Grundfos (Bjerringbro, Denmark) was considered here [15]. All pump models from single stage through sixteen-stage are assumed to be available in this series of products. The eight coefficients of Equations (3) and (4) for the single stage are obtained from regression analysis of the performance data as shown in Table 1.

The McQuay GRW360 model (McQuary, Minneapolis, MN, USA) which uses R407a refrigerant and has a nominal capacity of 30 refrigeration tons is used for the water-to-water heat pump [16]. The twenty coefficients from Equations (6) to (9) are calculated from regression analysis of the manufacturer's data as shown in Table 2. The reference values used in Equations (6) to (9) and the head loss coefficients of Equation (5) are also given in Table 2. Figure 3 shows performance curves of the heat pump according to heat source temperature, which is based on the catalog data.

Table 1. Pump model coefficients for the single stage.

A_0	A_1	A_2	A_3
1.022×10	-1.351×10^{-2}	2.5313×10^{-5}	-1.1589×10^{-7}
B_0	B_1	B_2	B_3
2.07403×10^{-1}	8.9077×10^{-4}	3.467×10^{-6}	-9.95602×10^{-9}

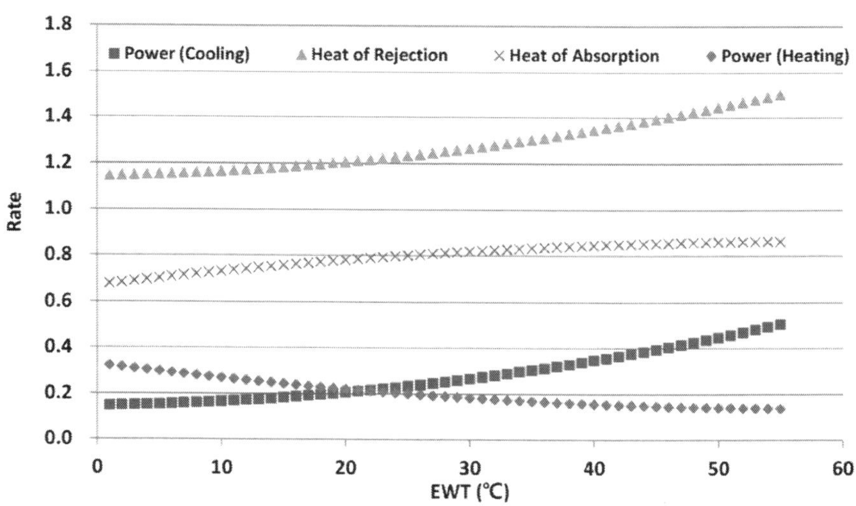

Figure 3. Performance curve of heat pump.

The head loss that occurs in the piping of the groundwater loop is approximated by Equation (10), and the head loss coefficient in the piping H_{piping} is determined to be 3.25×10^{-5}. The submersible pumps installed in the wells supply the groundwater with the head to cover the system head of the groundwater loop, which includes the pumping water level and the head losses in the heat exchanger and piping. The system head equation is shown in Equation (15):

Table 2. Heat pump model coefficients.

D_0	D_1	D_2	D_3	D_4
−3.70415	−2.26533	6.75804	3.322×10^{-2}	7.447×10^{-2}
E_0	E_1	E_2	E_3	E_4
−6.28243	7.83147	-6.663×10^{-1}	9.162×10^{-2}	3.253×10^{-2}
F_0	F_1	F_2	F_3	F_4
−6.42	5.66874	1.32662	-1.106×10^{-1}	1.483×10^{-2}
G_0	G_1	G_2	G_3	G_4
−8.35467	9.0202×10^{-1}	7.81833	2.878×10^{-2}	6.669×10^{-2}
$Q_{c,ref}$	$Q_{h,ref}$	$W_{c,ref}$	$W_{h,ref}$	C_{hp}
70 kW	70 kW	15 kW	15 kW	5.08245×10^{-5}
T_{ref}	$T_{S,in}$	$T_{L,in}$	V_{ref}	V_S
273 K	288 K	313 K	284 LPM	250 LPM

$$H_{SYS} = H_{GW} + H_{hp} + H_{hx} + H_{piping} \tag{15}$$

The head of the submersible pump and the system head vary as the groundwater flow rate varies. The residual head of the groundwater loop is defined as the difference between the groundwater loop system head and the submersible pump head. At the operating flow rate, the submersible pump head H_p matches the system head of the groundwater loop, meaning that the residual head becomes zero:

$$H_{res} = H_{SYS} - H_p \tag{16}$$

The Newton-Raphson method is used to find the operating flow rate. Iterations on the groundwater loop system head in terms of the groundwater flow rate are performed until the residual head vanishes. In the secondary loop, the chosen circulation pump is a Grundfos CR 15 single stage model (Manufacturer, Bjerringbro, Denmark). The operating circulation water flow rate of the circulation pump is 300 LPM with a head of 17.4 m. The power consumption of the pump is 1.17 kW, based on the performance data from the manufacturer's catalog.

5. RESULTS

The performance of the GWHP system can be expressed in terms of system COPs, which are the ratios of the heating or cooling capacities to the corresponding electricity consumptions of the GWHP system. Meanwhile, the COPs of the heat pump units are the ratios of the heating and cooling capacities to the corresponding electricity consumption for only the heat pump unit, excluding the pumps. System COPs are always lower than heat pump unit COPs, since the denominators of the ratios always become larger with the additional pump power consumptions. In calculating the GWHP system performance, the power consumption of the AHUs or FCUs and the pump in the load side loop are not included. The performance and characteristics of the components in the load side loop varies greatly for various types of building uses and designs. To evaluate the system performance of the GWHP without the effects caused by load side designs, the load side loop is assumed unchanged in the analysis.

5.1. Groundwater Level

The pumping groundwater levels are assumed to be obtained from a series of pumping tests. Pumping tests are experimental tests for finding the aquifer transmissivity or the soil permeability. In the test, a well is pumped at a controlled rate, and water-level response (drawdown) is measured in one or more surrounding observation wells. The response data from pumping tests is used to estimate the hydraulic properties of aquifers and evaluate well performance. The drawdown is a function of various parameters including pumping flow rate, aquifer transmissivity, time since the start of pumping, and distance from the well [17]. However, the water levels in the well during pumping tests include the drawdown effects, which are thus not being treated separately in this study.

System COPs of GWHP systems are presented for three different constant pumping groundwater levels: 15, 25 and 35 m. At each groundwater level, groundwater flow rates for the selected submersible pumps are found, and performance results of the GWHP system are calculated for the flow rates. The submersible pumps of eleven different stages are considered at each groundwater level. Performance results include heat pump COP and GWHP system COP.

System heating and cooling COPs are shown in Figure 4 and Figure 5, respectively. In Figure 4, as the groundwater level drops, the optimum flow rate for the GWHP heating system COP becomes more recognizable. In Figure 5, as the groundwater level drops, the slope of the curve in the low flow rate region becomes smaller in magnitude. However, in the high flow rate region, the system heating and cooling COPs drop rapidly as the flow rate increases.

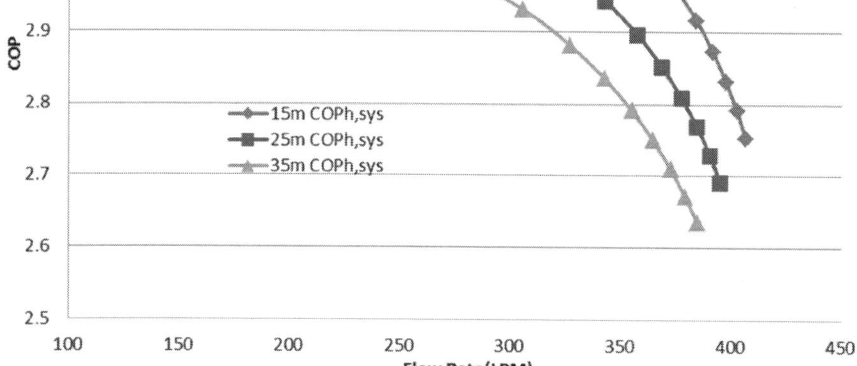

Figure 4. Heating system COP variation with the flow rate.

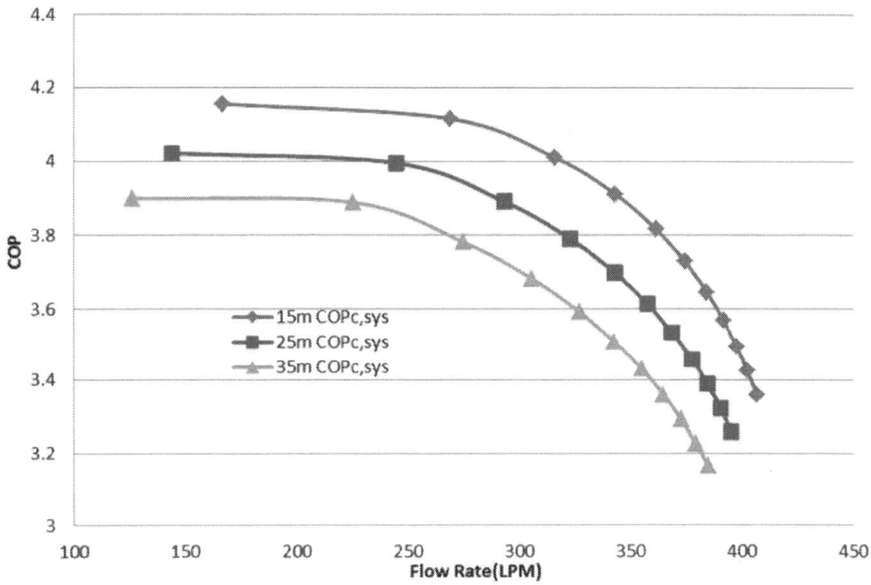

Figure 5. Cooling system COP variation with the flow rate.

The percentage of reduction of the COP of the GWHP system compared with that of the heat pump unit increases as the flow rate increases or the groundwater level drops. Figure 6 shows the system heating COP value variation with flow rates and groundwater levels. A 10%–15% reduction is shown for the 250 LPM groundwater flow rate. When the groundwater level is 35 m below the ground, the system heating COP shows the largest reduction (15%) as compared to the heat pump COP. At the higher flow rate of around 350 LPM, the system heating COP decreases by 15%–22% from the heat pump unit COP.

Figure 7 shows that the reduction percentage of system cooling COP as compared to the heat pump unit COP is larger than the reduction percentage of the system heating COP. A similar trend is found with the variation of flow rate and groundwater level. A 15%–22% reduction is observed for the 250 LPM groundwater flow rate. When the groundwater level is 15 m below the ground, the system heating COP shows the smallest reduction (15%) as compared to the heat pump COP. At the higher flow rate of around 350 LPM, the system heating COP shows a 22%–29% reduction from the heat pump unit COP.

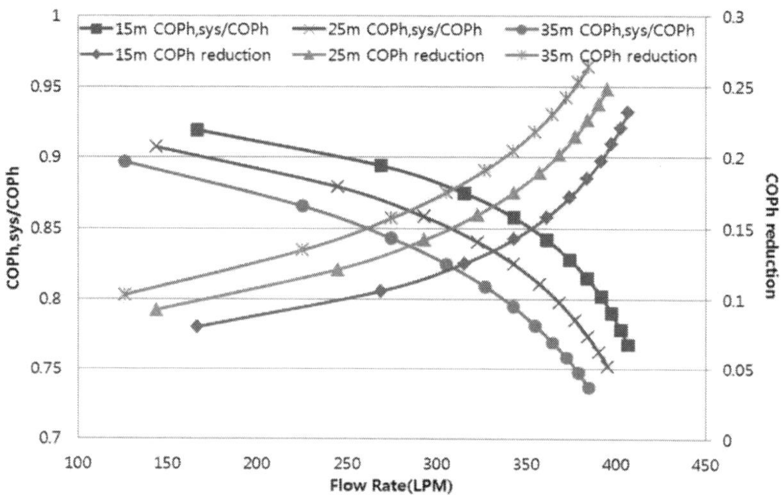

Figure 6. System heating COP reduction compared to heat pump unit COP.

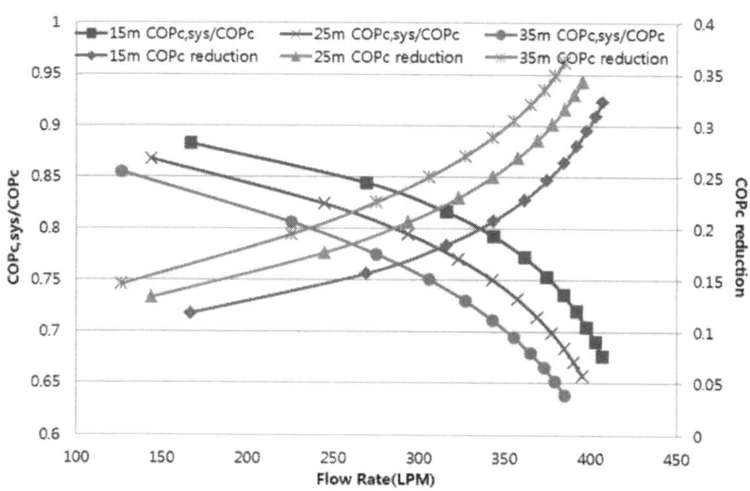

Figure 7. System cooling COP reduction compared to heat pump unit COP.

5.2. Groundwater Temperature

Groundwater temperature influences the system COPs of the GWHP system, as the heat pump capacities and COPs are dependent on the groundwater temperature. As the groundwater temperature increases, the heating COPs also increase, but the cooling COPs decrease. Conversely, as the groundwater temperature decreases, the heating COPs decrease, but the cooling COPs increase. While studying the effect of groundwater temperature on the system COPs of the GWHP, the groundwater level is set at a constant 25 m. The system COPs are calculated and shown in Figure 8 for groundwater temperatures of 12, 15 and 18 °C.

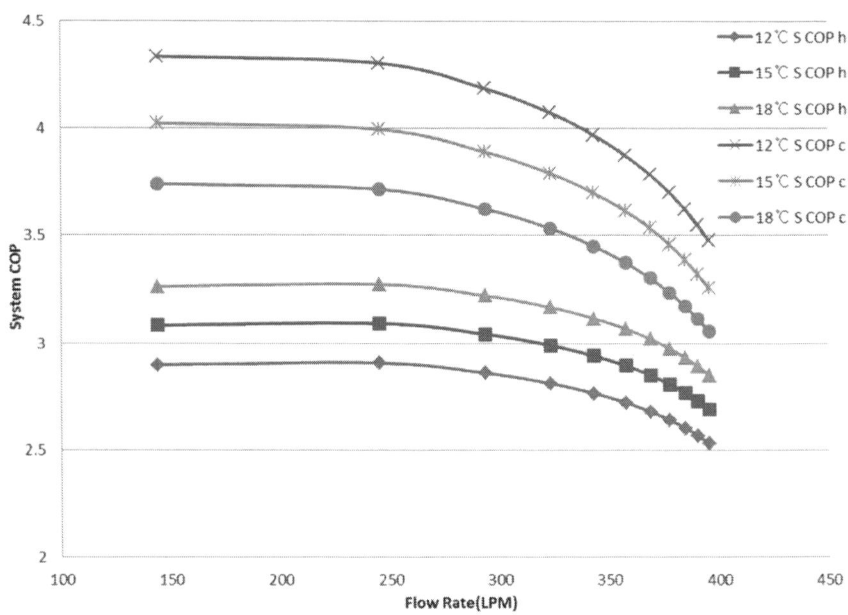

Figure 8. System Cooling and Heating COP variation to Groundwater Temperature.

The plot of the system cooling COPs for 12 °C groundwater shows the largest values; however, the system heating COPs show the smallest values. The system COPs for 18 °C groundwater show the opposite trend as compared to the case for 12 °C. The COPs for 15 °C groundwater lie between those for 12 and 18 °C. As the flow rate varies from 144.0 to 395.3 LPM, the system cooling COPs drop by approximately 19.0% (19.8% for 12 °C, 19.0% for 15 °C, and 18.3% for 18 °C), and the system heating COPs drop by approximately 12.7% (12.7% for 12 °C, 12.7% for 15 °C , and 12.6% for 18 °C).

At the flow rate of 245 LPM, the system heating COPs drop by 11.2% when compared with the COPs between 12 and 18 °C groundwater temperature, whereas the system cooling COPs drop by 13.7%. At a higher flow rate of 343 LPM, the system heating COPs drop by 11.2%, and the system cooling COPs drop by 13.1%

5.3. Heat Exchanger UA Value

Figure 9 shows the calculation results of system COP according to the change of heat exchanger UA values. The heat exchanger installed between the groundwater pump and the heat pump unit influences the system COP of the GWHP system. The temperature of the water circulated between the heat exchanger and the heat pump unit influence the capacity and power consumption of the heat pump unit.

The UA value of the heat exchanger determines the temperature differences between the hot and cold streams. In this study, a UA value of 12 W/K has been used for the previous results. A positive and negative variation by 25% from 12 W/K are considered, and three UA values of 9, 12 and 15 W/K are used to study its effect on the system COPs.

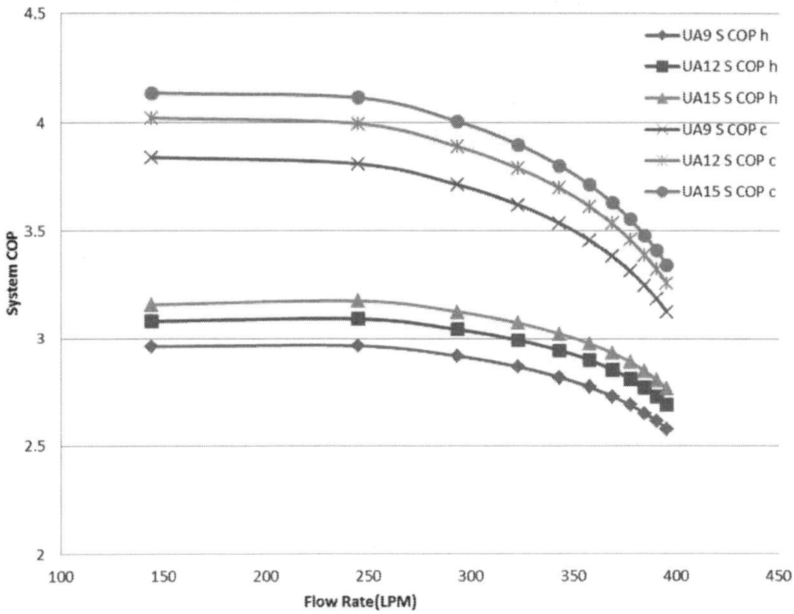

Figure 9. System cooling and heating COP variation to heat exchanger UA values.

When the UA value decreases from 12 to 9 W/K, the system heating COP drops by 4.2% and the system cooling COP drops by 4.6%. When the UA values increases from 12 to 15 W/K, the system heating COP rises by 2.7% and the system cooling COP rises by 3.0%. It is expected that the low UA values of the small heat exchanger may deteriorate the system COP values by a significant amount.

6. CONCLUSIONS

In this study, in order to develop an installation and operation guideline for a GWHP system, a numerical study was conducted using individual performance models and available system catalog data. The system COPs, as measures of performance for GWHP systems, were obtained by combining the component mathematical models. The system heating and cooling COPs were calculated for different groundwater levels, flow rates, groundwater temperatures, and UA values. The system COPs were almost unchanged or gradually changed in the low flow rate region as the flow rate increases. However, in the higher flow rate region, the system heating and cooling COPs decreased rapidly as the flow rate increased. The system heating COPs increased with warmer groundwater temperatures; similarly, the system cooling COPs increased with colder groundwater temperatures. Increasing the UA value of the heat exchanger universally improved the system heating and cooling COPs.

The system heating COPs were reduced 10%–15% at around 250 LPM and 15%–22% at around 350 LPM when compared to heat pump unit COPs. The system cooling COPs were reduced by a larger rate, 15%–22% at around 250 LPM, and 22%–29% at around 350 LPM. This result indicates that, if the flow rate is above 350 LPM at 30 refrigerant ton of heat pump, GWHP systems would be not proper due to low system COP.

At the flow rate of 245 LPM, the system heating COP dropped by 11.2% when 12 °C groundwater temperatures are compared to 18 °C, and the system cooling COP rose by 13.7%. At a higher flow rate of 343 LPM, the system heating COP dropped by 11.2%, and the system cooling COP rose by 13.1%. When the groundwater temperature rose, the system heating COP of GWHP systems increased, and the system cooling COP decreased. These results show that the effect of groundwater temperature on system performance is as significant as the effect of groundwater level or flow rate. Generally, the groundwater temperature at shallow depths is determined by the annual regional air temperature, but it could be different according to the well depth. Therefore, well designers should decide the well depth considering the groundwater temperature and building load, such as deeper wells for heating dominant buildings.

The system heating COP dropped by 4.2% and the system cooling COP dropped by 4.6% when comparing UA value of 9 to 12 W/K. When the UA value of 15 W/K is compared to a 12 W/K one, the system heating COP rose by 2.7% and the system cooling COP rose by 3.0%. Selection of proper UA values of the heat exchanger may ensure proper system COP values.

GWHP system design engineers need to find the groundwater flow rates at which GWHP systems maintain high system COPs with a proper selection of heat exchanger. The GWHP systems need to be modeled with detailed information on components, such as submersible pumps, heat pump units, and heat exchangers. In order to accurately predict system performance, proper GWHP system models need to be used in conjunction with the building heating and cooling load variations.

ACKNOWLEDGMENTS

This work was supported by the New and Renewable Energy (No. 20123040110010) of the Korea Institute of Energy Technology Evaluation and Planning (KETEP) grant funded by the Korea Government Ministry of Knowledge Economy.

AUTHOR CONTRIBUTIONS

All authors contributed equally to this work. All authors designed the simulations, discussed the results and implications and commented on the manuscript at all stages. Jinsang Kim performed the numerical simulations and the result analysis. Yujin Nam led the development of the paper and performed discussion.

CONFLICTS OF INTEREST

The authors declare no conflict of interest.

REFERENCES

1. Mendler, S. Buildings: An Essential Sector in Climate Change Mitigation; Civil Society Institute: Newton, MA, USA, 2012.
2. Bernstein, L.; Bosch, P.; Canziani, O.; Chen, Z.; Christ, R.; Davidson, O.; Hare, W.; Huq, S.; Karoly, D.; Kattsov, V.; et al. Intergovernmental Panel on Climate Change. In Climate Change 2007: Synthesis Report. Contribution of Working Groups I, II and III to the Fourth Assessment Report of the Intergovernmental Panel on Climate Change; Core Writing Team, Pachauri, R.K., Reisinger, A., Eds.; IPCC: Geneva, Switzerland, 2007.
3. Kwon, K.-S.; Lee, J.-Y.; Mok, J.-K. Update of current status on ground source heat pumps in Korea (2008–2011). J. Geol. Soc. Korea **2012**, 48, 193–199.
4. Park, Y.; Kwon, K.-S.; Kim, N.; Lee, J.-Y.; Yoon, J.-G. Change of geochemical properties of groundwater by use of open loop geothermal cooling and heating system. J. Geol. Soc. Korea **2013**, 49, 289–296.
5. Korea Energy Agency. New and Renewable Energy Center of Korea, Report of Geothermal Energy. Available online: https://www.renewableenergy.or.kr/ (accessed on 18 December 2015).
6. Korea Energy Agency. New and Renewable Energy Center of Korea, Legislation 2015-153. Available online: http://www.knrec.or.kr/knrec/14/KNREC140110.asp?idx=505&page=1&num=68&Search2=&Search=&SearchString= (accessed on 18 December 2015).
7. Kavanaugh, S.P.; Rafferty, K. Ground Source Heat Pumps, Design of Geothermal Systems for Commercial and Institutional Buildings; American Society of Heating Refrigeration and Air-Conditioning Engineers: Atlanta, GA, USA, 1997.

8. Nam, Y.; Ooka, R. Numerical simulation of ground heat and water transfer for groundwater heat pump system based on real-scale experiment. Energy Build. **2010**, 42, 69–75.

9. Zhou, X.; Gao, Q.; Chen, X.; Yu, M.; Zhao, X. Numerically simulating the thermal behaviors in groundwater wells of groundwater heat pump. Energy **2013**, 61, 240–247.

10. Spitler, J.D.; Rees, S.J.; Deng, Z. R&D Studies Applied to Standing Column Well Design; ASHRAE Report 1119-RP; American Society of Heating Refrigeration and Air-Conditioning Engineers: Atlanta, GA, USA, 2002.

11. Banks, D. An Introduction to Thermogeology: Ground Source Heating and Cooling, 2nd ed.; John Wiley & Sons: Oxford, UK, 2012.

12. Sachs, H.M. Geology and Drilling Methods for Ground-Source Heat Pump Installations: An Introduction for Engineers; American Society of Heating, Refrigeration and Air-Conditioning Engineers: Atlanta, GA, USA, 2002.

13. Kim, J.; Nam, Y. Study on groundwater heat pump system performance with various groundwater level. In Proceedings of the 11th IEA Heat Pump Conference, Montreal, QC, Canada, 12–16 May 2014.

14. EnergyPlus Engineering Reference: The Reference to EnergyPlus Calculations; U.S. Department of Energy: Washington, DC, USA, 2012.

15. Grundfos Product Guide: SP Submersible Pumps, Motors, and Accessories; L-SP-PG-001; Grundfos: Bjerringbro, Denmark, 2012.

16. Water to Water Source Heat Pumps 3 to 35 Tons R-22 and R407C Refrigerant/60 Hz; Catalog1111-1; McQuary: Minneapolis, MN, USA, 2007.

17. Park, B.-H.; Bae, G.-O.; Lee, K.-K. Importance of thermal dispersivity in designing groundwater heat pump (GWHP) system: Field and numerical study. Renew. Energy **2015**, 83, 270–279.

7

CHAPTER

ANALYSIS OF A VERTICAL FLAT HEAT PIPE USING POTASSIUM WORKING FLUID AND A WICK OF COMPRESSED NICKEL FOAM

Geir Hansen *, Erling Næss and Kolbeinn Kristjansson

Department of Energy and Process Engineering, Norwegian University of Science and Technology, Kolbjørn Hejes vei 1a, Trondheim 7491, Norway

ABSTRACT

Heat at high temperatures, in this work 400–650 °C, can be recovered by use of cooling panels/heat pipes in the walls of aluminum electrolysis cells. For this application a flat vertical heat pipe for heat transfer from a unilateral heat source was analyzed theoretically and in the laboratory, with special emphasis on the performance of the wick. In this heat pipe a wick of compressed nickel foam covered only the evaporator surface, and potassium was used as the working fluid. The magnitudes of key thermal resistances were estimated analytically and compared. Operating temperatures and wick performance limits obtained experimentally were compared to predictions. Thermal deformation due to unilateral heat flux was analyzed by the use of COMSOL Multiphysics®. The consequences of hot spots at different locations on the wick were analyzed by use of a numerical 2D model. A vertical rectangular wick was shown to be most vulnerable to hot spots at the upper corners.

Keywords: aluminum electrolysis cell; heat pipe; compressed nickel foam wick; potassium; hot spot analysis

1. INTRODUCTION

The current study originated from an industrial demand for heat recovery from electrolysis cell side-linings at temperatures in the range 400–650 °C. In aluminum electrolysis, about 50% of the total supplied energy is currently lost as heat to the surroundings. About 35% of the total heat loss is through the walls of the electrolysis cells [1], which corresponds to ca. 2.4 kWh/kg of produced aluminum. Cooling panels in the side walls of the electrolysis cells were proposed by Aune et al. [2] for improved process control, as well as for high temperature heat recovery. The panels were specified to utilize evaporation and condensation of a working fluid and the recovered heat can be used for, for instance, electricity production. The temperature of the recovered heat depends on the position of the cooling panels in the side-linings of the cells; in the current study, a position corresponding to a temperature range 400–650 °C was specified. The main components of the electrolyte bath are cryolite (Na_3AlF_6) and alumina (Al_2O_3) [1]. The melting point of aluminum is 660 °C, but the cryolite (in which the alumina is dissolved) requires a temperature of 1010 °C for melting [1]. The electrolyte bath is very corrosive, but with a working temperature of 400–650 °C in the cooling panels it is possible to create and maintain a protective frozen ledge on the inside of the cell walls. Figure 1 shows a sketch of the cross section of an electrolysis cell equipped with a wall cooling panel [2] (only a part of the cell is shown). The captions in Figure 1 are listed in Table 1. The protective frozen side ledge (24) is formed on the cooling panel. The panel can be exposed to hot spots if the protective frozen side ledge breaks down. In such cases, the hot molten electrolyte (12) can come directly in contact with the cooling panel evaporator (8), through cracks in the frozen side ledge. The cooling panel represents the first stage in a heat recovery system, in which heat is transferred from the cooling panel evaporator (8) via the condenser (9) to a cooling circuit (10C) by evaporation and condensation of a working fluid.

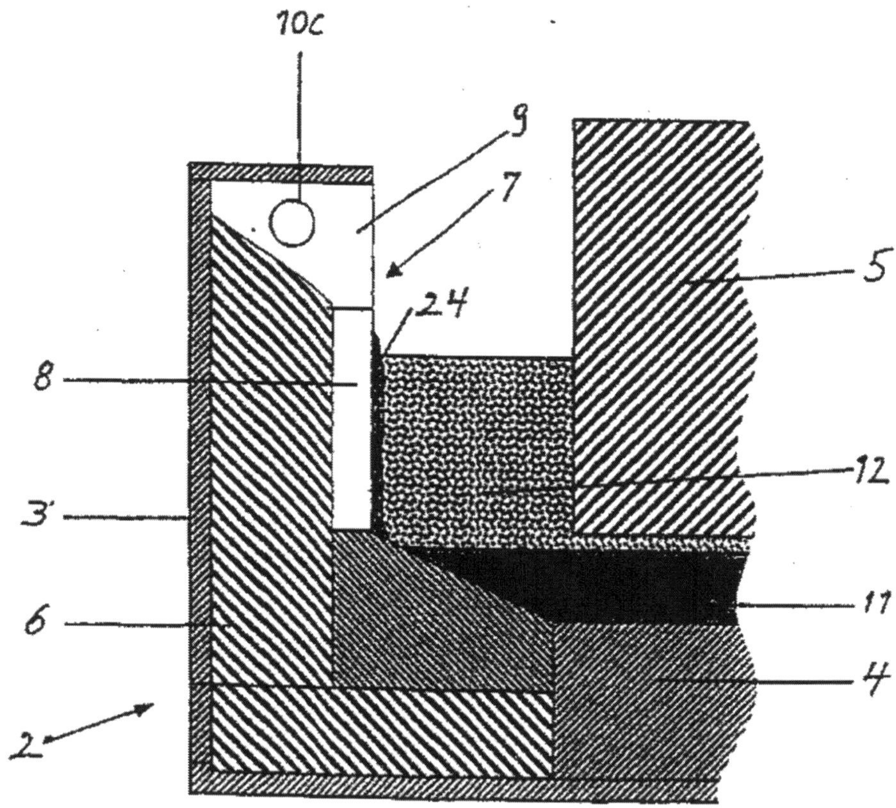

Figure 1. A cross-section of an electrolysis cell with cooling panel [2].

Table 1. Captions in Figure 1.

Number	Explanation
2	Electrolysis cell
3	Steel shell
4	Cathode
5	Anode
6	Insulation
7	Cooling panel
8	Cooling panel evaporator
9	Cooling panel condenser
10C	Cooling circuit
11	Aluminum
12	Electrolyte bath of alumina (Al_2O_3) and cryolite (Na_3AlF_6)
24	Solid electrolyte (frozen side ledge)

The current paper reports:

- A vertical flat heat pipe considered for use as cooling panel;
- Experimental results obtained for the built heat pipe, in a laboratory test rig;
- A thermal resistance network for the built heat pipe;
- Results from an analytical spreadsheet model of the built heat pipe;
- A numerical study of the thermal expansion of the heat pipe, using finite-element (FEM) analysis; and
- A numerical hot spot analysis, undertaken to determine the most vulnerable position(s) of hot spots on a vertical flat heat pipe wick.

Heat pipes of many forms and for many applications have been proposed and are analyzed in textbooks (for instance, [3,4,5,6]); a review including recent advances was given recently by Faghri [7]. In some areas, for instance computer cooling, heat pipe technology is well established; however, due to new areas of application for heat pipes the components, materials, and physical processes of heat pipes are still receiving considerable attention. Reay and Harvey [8] recently reviewed the role of heat pipes in intensified unit operations, and pointed specifically at chemical reactors as highly relevant application areas. Heat pipe technology is in the current study considered for the cooling panels of the aluminum electrolysis cells.

In traditional "capillary heat pipes" the inner surfaces of the evaporator, the condenser, and the adiabatic section are all covered by a wick material, or have grooves serving as capillary pumps, wetting the surface. The capillary pressure is the driving force for the flow and distribution of the working fluid in the wick/groove, and in many cases the performance of the wick/groove will be the main limiting factor of the heat pipe. Wicks are still being developed and improved in order to best fit the application requirements. For instance, advanced bimodal wicks have been designed by sintering fine wick structures to coarse wick structures [9]. Another advanced wick has a graded structure which reflects the pressure gradient inside the heat pipe [10]. Several references on wicks of sintered, groove and mesh type were listed in a recent review by Chan et al. [11]. Faghri [3] divided wicks into two groups; homogeneous and composite, and stated that homogeneous wicks are usually chosen due to simple design, manufacture, and installation (i.e., lower costs), even though the composite wicks usually have better performance. Wicks can be made of compressed nickel foam, and the development and testing of such wicks were reported in an earlier paper [12]. This type of wick has the advantage that the wick properties can be selected within a wide range by choosing the degree of compression [13].

Heat pipe wicks are characterized by their porosity, permeability, thermal conductivity, and effective pore radius with regard to a specific fluid. The wick porosity and permeability are properties of the pore structure only [14], independent of the fluid applied. The effective pore radius depends on the contact angle between the wick structure and the fluid applied and is, therefore, not fluid independent. The porosity is usually determined according to a standardized procedure [15]. Various methods exist for the determination of the permeability and the effective pore radius, and in most cases these properties are determined by use of a model fluid and without exposing the wick to a heat flux, for instance as in [16]. Shirazy and Fréchette [17] recently questioned the validity of the effective pore radius determined this way for the calculation of the wick heat transfer capacity (the capillary limit). They had experienced that wicks made of copper foam performed much better in heat pipes than predicted from existing correlations for the capillary limit. Their explanation was that the effective pore radius will decrease during evaporation due to meniscus recession, and thereby increase the wick heat transfer capacity (compared to the value calculated from an effective pore radius determined from an experiment in which the wick was not exposed to heat). More studies of this hypothesis and comparative studies between the different methods for the determination of the permeability and the effective pore radius will increase the reliability of wick heat transfer capacity predictions.

With regard to working fluid, attention is here given to the alkali metal potassium, which is a commonly-used working fluid in traditional heat pipes for the temperature range 500–1000 °C [4,7]. Alkali metals are very reactive, but excellent working fluids in heat pipes when treated correctly. Potassium has to be handled in a high-quality glove box filled with inert gas (such as Ar), or under vacuum, due to the reactivity with oxygen and water (moisture). Hydrogen gas will be produced in the reaction between potassium and water. Potassium will also tend to remove oxide from the surface of steel [18]. Chemical compatibility between the envelope material and the working fluid is essential in order to avoid corrosion and/or inert gas generation. Ten years of successful operation was reported for a heat pipe made of stainless steel 718 envelope and 316L wick material, utilizing sodium as a working fluid at 750 °C operating temperature [19]. In industry, lifetimes of years are usually needed and expected for heat pipes, often with harsh thermal, chemical, and mechanical loads. Compatibility between the heat pipe and its surroundings is, therefore, equally important to the internal compatibility between the envelope, wick, and working fluid. Metals are the dominating envelope and wick materials, but high temperature wickless heat pipes (thermosyphons) of ceramic material are also being developed; results from promising experiments with sodium and zinc as working fluids were reported recently [20]. Nickel is compatible with potassium [4], but starts oxidizing rapidly at about 350 °C [21] and a protective/reducing gas atmosphere of 5 %(v) H_2, 95 %(v) Ar is therefore often used during the heat treatments and preparations of the nickel wicks and heat pipe envelopes. Stainless steel 304 is also reported to be compatible with potassium [4], and may be a more economical choice than nickel for heat pipes for industrial applications.

Panels/heat pipes in the side lining of the aluminum electrolysis cells should be as slim as possible. Attention must, therefore, be given to the sonic limit for the vapor flow, especially when alkali metal working fluids are applied, due to their low vapor densities [22]. The entrainment and flooding limits will also be important if the vapor and condensate are in direct contact and flowing in opposite directions in a heat pipe/thermosyphon.

2. EXPERIMENTAL CHARACTERIZATION

An experimental study was carried out to investigate the performance of a flat vertical heat pipe under temperature and heat flux conditions expected in the walls of aluminum electrolysis cells (including hot spot heat fluxes). The hypothesis was that a wick in the evaporator section would be a robust solution for the distribution of working fluid over the evaporator surface under varying temperature and heat flux conditions.

The Heat Pipe

An experimental heat pipe was made specifically for this study. The built experimental heat pipe is shown in Figure 2, to the left fully assembled, to the right split in two.

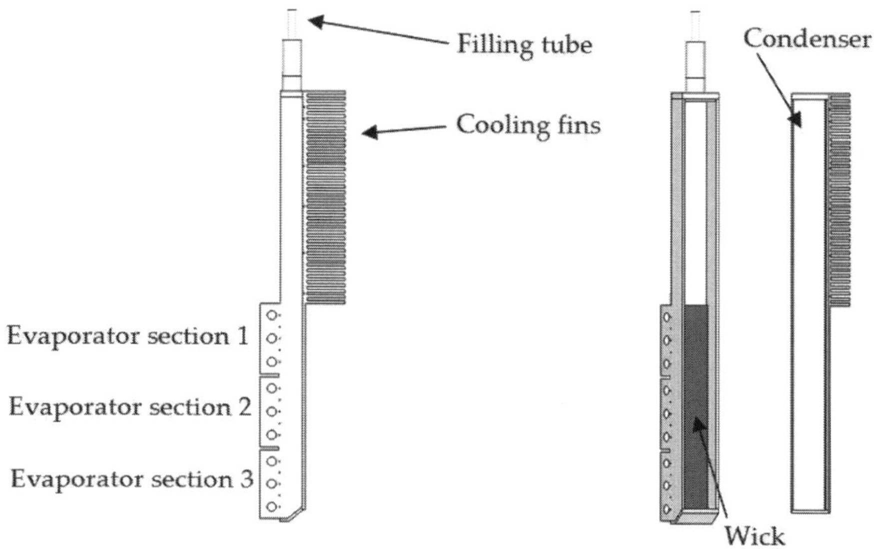

Figure 2. The experimental heat pipe fully assembled (**left**) and split in two (**right**).

For the current experimental study, nickel was selected both for the envelope and the wick, due to its relatively high thermal conductivity and reported compatibility with potassium. The heat pipe was designed "as slim as possible", according to the requirements for the cooling panel for electrolysis cell walls. However, the dimensions of the flow channel were chosen such that the sonic limit, the entrainment limit, and the flooding limit were unlikely to cause problems. Correlations available in reference [3,4] for predicting these limits were used in the design process. The flow channel width was chosen so large that the sidewalls of the heat pipe would likely not cause significant disturbance of the fluid flow.

The experimental heat pipe had no adiabatic section, due to manufacturing limitations. The total height of the evaporator was 150 mm. The inner height of the condenser section was 142 mm. The inner width of the flow channel was 25 mm and the depth 11.55 mm. A wick, 25 mm × 150 mm × 1.15 mm, consisting of four layers of compressed and sintered nickel foam was attached to the evaporator surface by use of electron beam spot welding. The evaporator was divided in three sections where each had three cartridge heaters. The motivation for having multiple sections was to be able to impose a non-uniform heat flux distributions over the evaporator surface. The cartridge heaters were mounted into the heat pipe having force fit tolerances, hence no thermal paste was necessary. The supplied power to each evaporator section was controlled by phase angle firing thyristors. Each cartridge heater was rated 225 W at 240 V, and had a maximum work temperature of 760 °C. For an ideal case without heat losses the nine cartridge heaters could, from their specifications, provide a maximum uniform heat flux of 540 kW/m^2. The temperatures in the evaporator wall were measured by 3 × 5 type N thermocouples, Ø1 mm, installed in pockets in each section of the evaporator wall. Similar thermocouples were also installed in pockets in the condenser wall and on the outside of the heat pipe adiabatic walls (the walls opposite of the evaporator, respective condenser surfaces). LabView® together with modules NI 9214, NI 9219, NI 9225, and NI 9227 from National Instruments® were used for data recording and parameter control. The uncertainty of the measurement of the power supplied to the cartridge heaters was less than 1%. The thermocouples attached to the logging system were calibrated by use of a JOFRA RTC-700 calibrator (accuracy ± 0.11 K). The amount of potassium filled into the heat pipe was 10.2 g, which was estimated to be sufficient for full wetting of the wick, with a good safety margin, for all planned experimental operating conditions.

The outside of the condenser section was equipped with fins and was cooled by nitrogen gas. The heat taken up by the nitrogen gas was calculated from the measured nitrogen flow rate and the temperature increase of the nitrogen over the condenser ($QN2{=}m^{\cdot}N2{\cdot}cp{,}N2{\cdot}\Delta TN2$). The inlet temperature of the nitrogen was kept at 50 °C. The nitrogen flow rate was determined from the volume flow rate measured by use of a Sierra 620S insertion flow meter in the cooling circuit. The range of the Sierra flow meter was 0–23.45 Nm^3/h and the uncertainty ± 0.28% = 0.066 Nm^3/h, where N in this particular case was defined as 15 °C and 1 atm. The heat loss to the surroundings was the difference between the heat supplied to the heat pipe by the cartridge heaters and the heat taken up by the nitrogen gas. For safety reasons, the heat pipe was in a container filled with nitrogen gas during the experiments.

3. THEORETICAL ANALYSIS

3.1. Wick Performance

The limitation that directly addresses the performance of the wick alone is defined by the pressure balance [3]:

$$\Delta P_{cap,\max} \geqslant \Delta P_{hs} + \Delta P_f \tag{1}$$

The pressure loss of the vapor flow in the heat pipe is not taken into account here, as attention is given to the wick only. Inertia effects in the wick and the pressure losses associated with the interfacial thermal resistances are also omitted because these terms are small compared to the terms of Equation (1) in the current case.

If the tortuous flow paths in the wick are represented by a capillary tube model then the capillary pressure can be expressed from the Young–Laplace equation as:

$$\Delta P_{cap,\max} = \frac{2\sigma}{r_{eff}} \tag{2}$$

The hydrostatic pressure of the working fluid in the wick relative to the vapor pressure outside the wick is:

$$\Delta P_{hs} = \rho_l g H \tag{3}$$

The frictional pressure loss due to the flow of working fluid in the wick is calculated by use of Darcy's law:

$$\Delta P_f = \frac{\mu_l}{\kappa A_c \rho_l} \int_0^H \dot{m}_l \, dy \tag{4}$$

The total heat transfer capacity (in Watts) of a vertical flat wick exposed to a uniform heat flux is obtained by inserting Equations (2)–(4) into Equation (1) [12]:

$$Q = \dot{m}_{l,\max} \cdot h_{fg} = \frac{2\kappa A_c \rho_l}{H \mu_l} \left(\frac{2\sigma}{r_{eff}} - \rho_l g H \right) h_{fg} \tag{5}$$

Equation (5) has been applied to predict the heat transfer capacity of the wick in the capillary pressure-limited range. In the vapor static pressure-limited range the capillary pressure term ($2\sigma/r_{eff}$) in Equation (5) is replaced by the vapor pressure of the working fluid [21].

3.2. The Numerical Hot Spot Analysis

A numerical 2D hot spot analysis for a flat vertical wick was carried out, considering the combined effect of frictional and static pressure losses due to localized regions of high heat flux. The method is quite similar to the method described in [23]. The relevant equations were discretized and solved numerically using MatLab®.

Starting with the continuity equation:

$$\frac{\partial u}{\partial x} + \frac{\partial v}{\partial y} = \frac{\alpha}{\rho} \tag{6}$$

where $\alpha = f(x,y)$ is a mass flux source/sink function, and expressing the velocity components *u* and *v* respectively using Darcy's law:

$$u = -\frac{\kappa}{\mu}\frac{\partial P}{\partial x} \quad ; \quad v = -\frac{\kappa}{\mu}\frac{\partial P}{\partial y} \tag{7}$$

Equation (7), inserted into the continuity equation, Equation (6), yields Equation (8) which describes the pressure field due to flow friction in the wick:

$$\frac{\partial^2 P}{\partial x^2} + \frac{\partial^2 P}{\partial y^2} = -\frac{\mu\alpha}{\rho\kappa} \tag{8}$$

Appropriate boundary conditions are:

Through the left and right boundaries there is no flow, i.e.,:

$$\left(\frac{\partial P}{\partial x}\right)_{right,left\, boundary} = 0 \tag{9}$$

No flow through the top boundary, i.e.,:

$$\left(\frac{\partial P}{\partial y}\right)_{top\, boundary} = 0 \tag{10}$$

At the bottom of the wick the following boundary condition applies:

$$P_{bottom} = P_{sat} - \rho g H \tag{11}$$

where the hydrostatic pressure as a function of height from the liquid bath, y, is:

$$P_{hydrostatic} = \rho g y \tag{12}$$

H is the height of the wick. It should be emphasized that this definition of P_{bottom} is a fictive pressure, created as an intermediate step in the calculation, the real pressure at the bottom of the wick is P_{sat}. The boundary condition P_{bottom} as defined by Equation (11) may be seen as the pressure available to overcome flow friction.

The total pressure field in the vertical wick is:

$$P_{total}\,(x,y) = P_{flow\,friction}\,(x,y) + P_{hydrostatic}\,(y) \tag{13}$$

The total pressure is compared to a critical pressure, which may be seen as the "pressure remaining in the wick" and has minimum values defined by Equations (14) and (15):

$$P_{critical} = \left(P_{sat} - \frac{2\sigma}{r_{eff}}\right)\ when\ P_{sat} \rangle \frac{2\sigma}{r_{eff}} \tag{14}$$

$$P_{critical} = 0\ when\ P_{sat} \langle \frac{2\sigma}{r_{eff}} \tag{15}$$

The calculation procedure is as follows:

1. The pressure field in the wick is calculated from Equation (8) and the boundary conditions, i.e., the pressure field is determined entirely by the flow friction, starting from the saturation pressure minus the hydrostatic pressure at the bottom of the wick.

2. The total pressure field in the wick is then calculated by adding the hydrostatic pressure, as a function of height y, to the pressure field generated in point 1, according to Equation (13).

3. For the selected heat fluxes and hot spot the minimum pressure in the total pressure field is compared to the critical pressure defined by Equation (14) or Equation (15). The

wick works well if the minimum total pressure is higher than the critical pressure. The capacity of the wick has been exceeded if the minimum total pressure is lower than the critical pressure.

3.3. Thermal Resistance Model for the Whole Heat Pipe

The thermal resistance network for the heat pipe is shown in Figure 3. Heat is supplied from the cartridge heaters at the lower left of the resistance network. The heat leaves the network with the nitrogen gas to the right or as heat loss to the left.

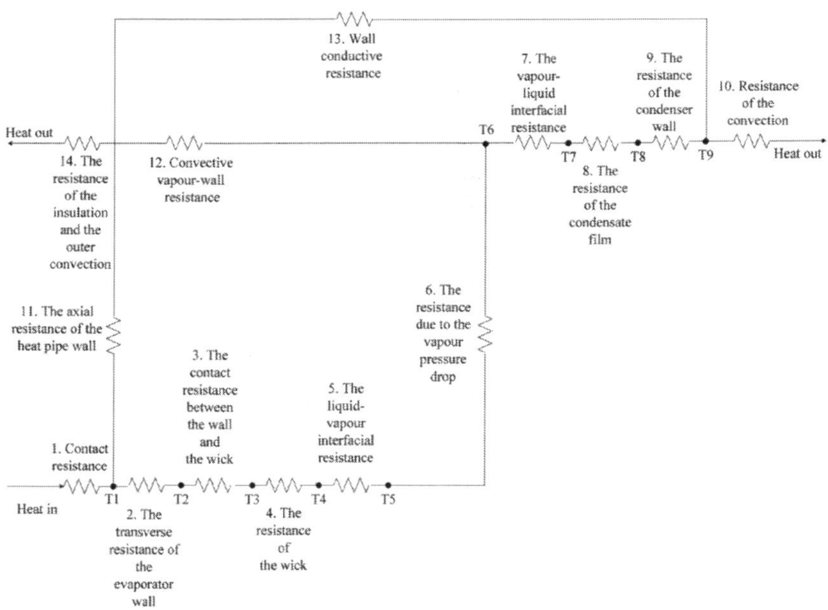

Figure 3. The thermal resistance network of the heat pipe.

A simplified model of the heat pipe based on major heat transfer resistances was developed and implemented in a spreadsheet. With known dimensions and boundary conditions, the individual

Table 2. Thermal resistances taken into account.

Thermal Resistance	Expression for R'' ((m^2·K)/W)	Reference
2. The transverse resistance of the evaporator wall.	$\frac{L}{k_{Ni201}}$	[24]
4. The resistance of the wick.	$\frac{2L}{\left[(1-\varepsilon)k_{solid}+\varepsilon k_{fluid}\right]+\left[\frac{k_{fluid}\cdot k_{solid}}{\varepsilon k_{solid}+k_{fluid}(1-\varepsilon)}\right]}$	The average of parallel and serial wick resistances [3] was assumed.
5. and 7. The liquid-vapor and vapor-liquid interfacial resistance.	$\frac{1}{\left(\frac{2\alpha}{2-\alpha}\right)\left(\frac{h_{fg}^2}{T_v v_{fg}}\right)\sqrt{\frac{M_v}{2\pi R_u T_v}}\left(1-\frac{p_v v_{fg}}{2h_{fg}}\right)}$	[3]
8. The resistance of the condensate film.	$\frac{\left(\frac{3\mu_l\Gamma}{g\rho_l(\rho_l-\rho_v)}\right)^{\left(\frac{1}{3}\right)}}{k_l}$	[24]
9. The resistance of the condenser wall.	$\frac{L_{wall}}{k_{Ni201}}$	[24]
10. Resistance of the nitrogen gas convective cooling.	$\frac{1}{h_{convection}}$	[24]

resistances in Table 2 were calculated. This process was iterative, as the operating condition in the heat pipe had to be determined. For simplicity, metal/metal thermal contact resistances were neglected, as were axial conduction in the heat pipe wall, resistance due to vapor pressure loss, and blowing and suction effects in the evaporator and condenser. The heat transfer coefficient for the nitrogen convective gas cooling, $h_{convection}$, was calculated for flow in a rectangular channel because, when mounted in the test rig, the flow areas between the heat pipe fins became rectangular channels. Fully developed flow was assumed, which gave a conservative estimate for $h_{convection}$. A fin efficiency was calculated to account for the temperature gradient in the fins. The magnitudes of the thermal resistances listed in Table 2 are compared in the Results and Discussion section. The spreadsheet model was also used to estimate the operating temperature (T_5 in Figure 3) of the heat pipe for the following input parameters: heat input, nitrogen gas flow rate, and inlet temperature. Operating temperatures calculated by use of the spreadsheet model have been compared to measured operating temperatures in the Results and Discussion section. With regard to uncertainty, the input parameters are based on the measurements and have the same uncertainties as them. The output parameters have, in addition, the uncertainties from the different resistances in the network plus the uncertainty of the simplifications, which are not quantified.

4. RESULTS AND DISCUSSION

4.1. Thermal Resistance Model

Using the developed model, the magnitudes of the heat pipe related thermal resistances were calculated and are compared in Figure 4 for an operating temperature of about 500 °C. The thermal resistance on the coolant side has been omitted in Figure 4 as this resistance depends on the cooling fluid (in this case N_2-gas) and is not related to the heat pipe characteristics (for the experimental heat pipe setup, the coolant-side resistance represented 85% of the total resistance due to the poor heat transfer characteristics of the nitrogen gas). Figure 4 shows the relative magnitudes of the resistances associated with the heat pipe, for 2 mm envelope wall thickness overall. The thermal resistances associated with the conduction through the evaporator and condenser walls were the largest, followed by the thermal resistance of the wick, the interfacial resistances, and finally the thermal resistance of the condensate film. The interfacial thermal resistance at the vapor/liquid interface was about twice the thermal resistance of the condensate film, which emphasizes that interfacial thermal resistances can be important in studies of metal vapor condensation [25]. The magnitude and relative importance of the interphase thermal resistances increase with decreasing temperature; however, at an operating temperature of 500 °C the thermal resistances associated with the condensation of the potassium constitute only a small part of the total thermal resistance. The total thermal resistance between the evaporator and condenser external surfaces is 0.031 K/W, representing an effective axial thermal conductivity of the heat pipe of 10,560 W/(m·K) for an effective length of 14.6 cm and an outer axial cross-sectional area of 451×10^{-6} m^2. For comparison, this represents a thermal conductivity of 26.4 times the thermal conductivity of copper (400 W/(m·K) [24]).

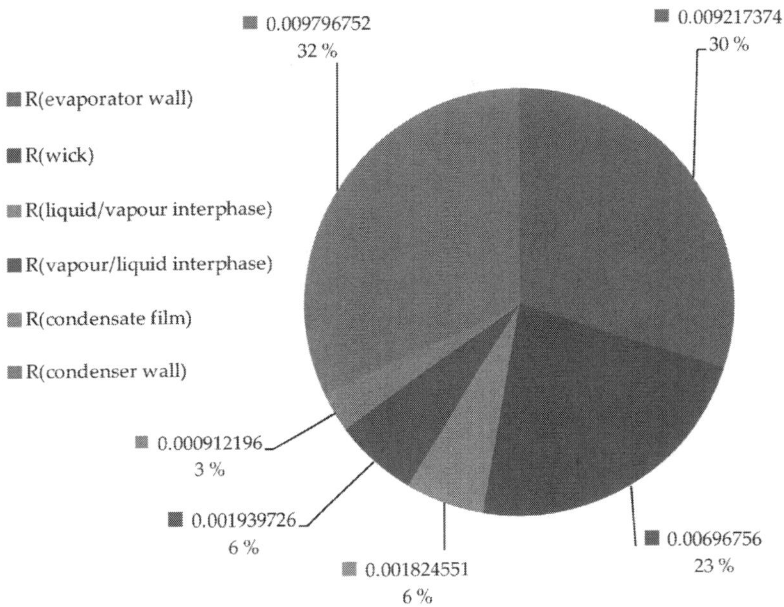

Figure 4. The importance of the considered thermal resistances [K/W] at $T_5 \approx 500$ °C.

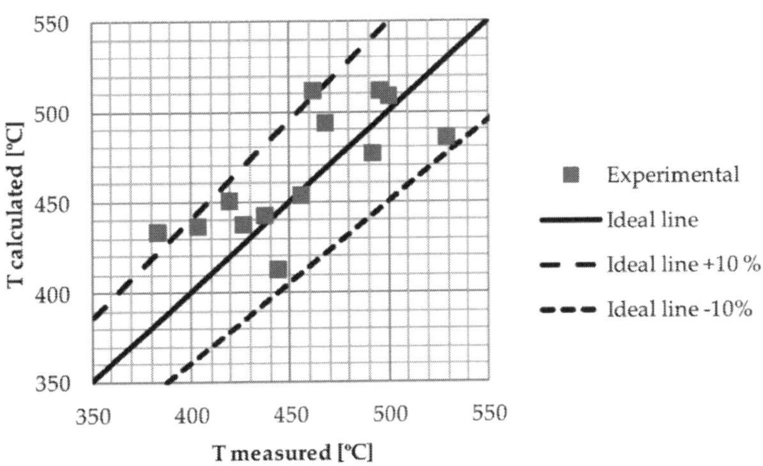

Figure 5. Calculated vs. measured operating temperatures, at position T_5 (Figure 3).

In Figure 5 the heat pipe operating temperatures calculated by use of the spreadsheet model are compared to measured operating temperatures.

For the limited set of data comparisons the simplified thermal resistance model seems to reproduce the experiments to within ±10%. The points in Figure 5 were obtained at a wide range of experimental conditions; heat fluxes were in the range from 38 kW/m^2 to 360 kW/m^2 and N_2 flow rates between 2.3 Nm^3/h and 17.2 Nm^3/h. The main reason for the prediction deviation is related to transient heat losses to the insulation and safety chamber not fully accounted for in the spreadsheet model.

4.2. Performance Limitations of the Wick, Analytical, and Experimental Results

The capillary pressure in a heat pipe wick can never be higher than the vapor pressure of the working fluid. This will limit the performance of the heat pipe wick, as it is the capillary pressure (or the vapor static pressure) that will be the "driving force" in Equation (5).

The active height of the heat pipe wick was 11.2 cm, and the wick capacity limitations are shown in Figure 6.

For a uniform heat flux the dryout at the exceedance of the wick maximum capacity would be expected to occur at the top of the wick, i.e., an accompanying temperature peak would be registered at the top of evaporator Section 1 in Figure 2. For the experiments reported in Figure 6 the heat flux was increased in small steps, and the temperatures in the evaporator were monitored graphically on the computer screen. For heat fluxes above the reported measured limits the rapid increase in temperature did not occur at the top of the wick, but at the top of evaporator Section 2. This could indicate that the evaporator surface dryout and the accompanying temperature peak were caused by a local contact problem between the wick and the evaporator surface at the top of evaporator Section 2, rather than by the exceedance of the "pumping" capacity of the

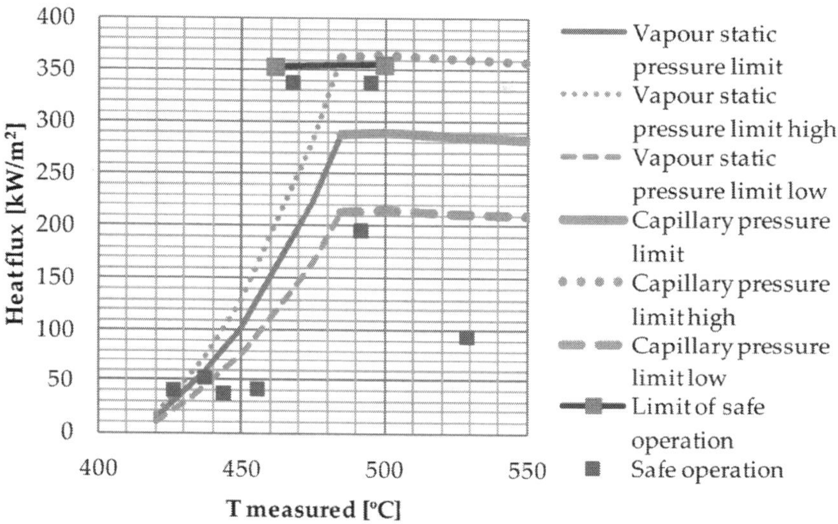

Figure 6. Wick performance data for wick height 11.2 cm. Dots represent experimental data. Red dots—at maximum performance limit; blue dots—safe operation.

wick. In such case, the maximum capacity of the wick could be even higher than shown by the experimental points in Figure 6. The most likely scenario is that such a contact problem was caused by thermal expansions, which are largest at the maximum heat flux. In the following discussion it is assumed that the measured maximum reported in Figure 6 represents the maximum capacity of the 11.2 cm wick, even though the dryout did not occur at the top.

From Figure 6 it is observed that the heat pipe was able to operate above the vapor static pressure limitation calculated for the whole effective wick height (11.2 cm), while in the capillary pressure limited range the performance overlapped the calculated limit when the experimental uncertainty was taken into account. The large uncertainty indicated in Figure 6 by the high and low vapor static pressure limit curves and the high and low capillary pressure limit curves is a result of the uncertainty of the permeability of the compressed wick, which was calculated to ±26% [12]. In general, from Equation (5), the heat transfer capacity will increase by decreased effective pore radius and/or increased permeability. Increased effective pore radius due to a heat flux, as described by Shirazy and Fréchette [17], will only have effect in the capillary pressure limited region. The experimental point above the vapor static pressure limit line can, therefore, only be explained if the real permeability of the wick was higher than the value (73×10^{-12} m^2) used in the calculation of the corresponding limit. However, the calculated limits are extremely sensitive to the permeability. Estimates have shown that a small channel in the wick, or between the heated wall and the wick, in the order of 34 μm, will increase the permeability sufficiently to explain the deviation between experiments and predictions. It is not unlikely that such a channel can have been created between the compressed foam layers of the wick due to the mechanical handling of the heat pipe, or due to thermal expansion, as discussed in the following.

4.3. Thermal Expansion

A heat transfer panel/heat pipe located in the side lining of an aluminum electrolysis cell will be exposed to unilateral heat fluxes. The temperature gradients caused by the unilateral heat flux will cause some thermal deformation for any real structure of finite thermal conductivity, such as the heat pipe envelope. The thermal deformation of the heat pipe was analyzed by use of Comsol Multiphysics®. Figure 7 shows the results from one of the most extreme cases, where about 480 kW/m^2 were applied to each of the upper two evaporator sections and about 160 kW/m^2 was applied to the lower evaporator section.

The heat pipe was fixed at the fin base. The results shown in Figure 7 indicate that the heat pipe bends more than 2 millimeters due to the unilateral heat flux, despite the relatively high thermal conductivity of the envelope material (Ni201). Under the current conditions the temperature at the cartridge heaters was about 90 K above the operating temperature of the heat pipe. Awareness about thermal deformation is important in future design processes in order to implement solutions which can prevent potential thermal contact problems on the outside of the heat pipe, as well as between the wick and the evaporator surface.

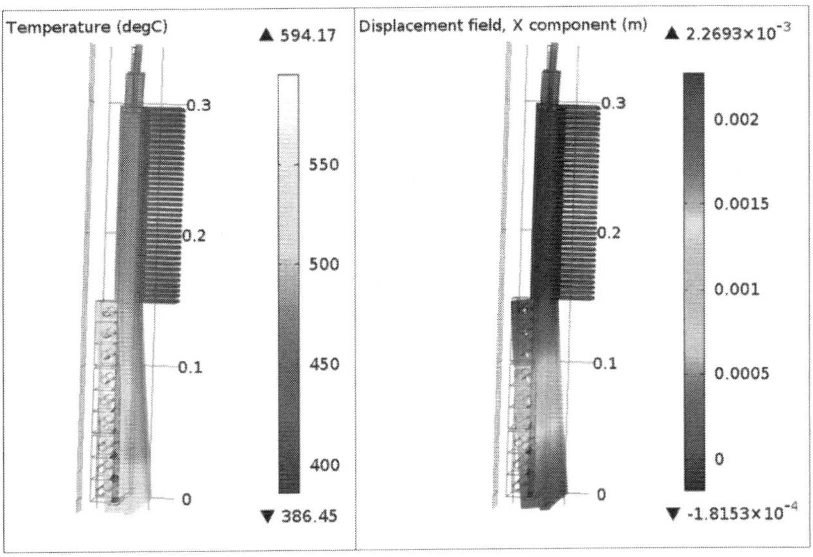

Figure 7. Temperature distribution [ºC] and corresponding thermal deformation [m] of the heat pipe.

4.4. Numerical 2D Hot Spot Analysis

In this section the results of the numerical study of hot spots on vertical flat wicks are reported. Different scenarios were considered, exemplified here by analysis of the consequences of hot spots in the different positions, 1 to 9, of the wick in Figure 8. The maximum uniform heat flux for the whole wick is used as reference in the discussion. Dryout will occur at the top of the wick if the maximum uniform heat flux is exceeded.

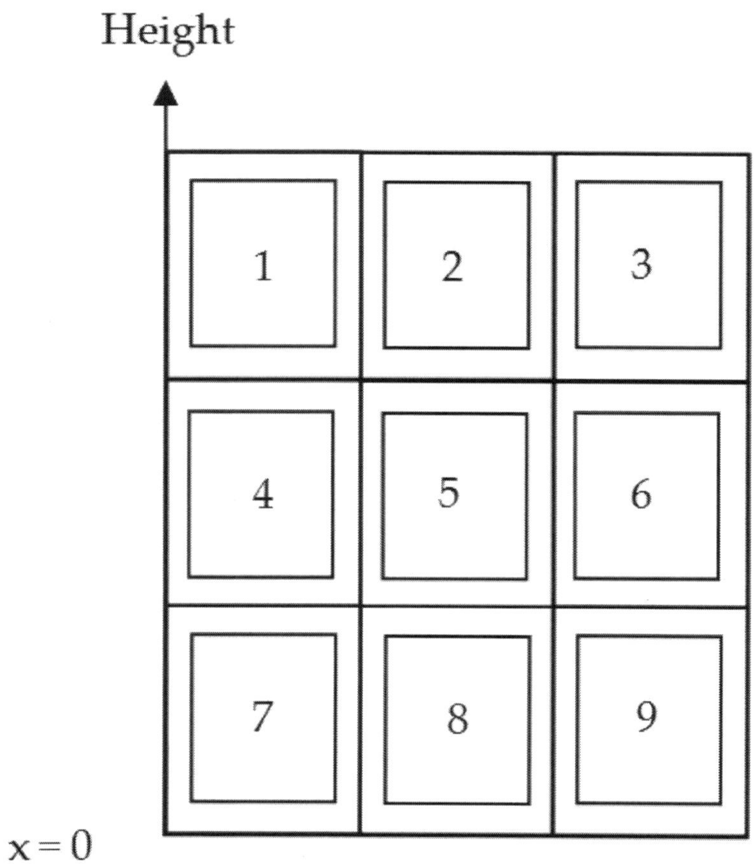

Figure 8. A vertical rectangular wick with nine potential hot spot positions.

The experimental heat pipe wick (0.025 m wide) represented just a slice of a larger wick for a cooling panel. The width of the wick in the numerical study was set to 0.112 m, giving a quadratic wick for this analysis. The size of the hot spot was 0.0112 m × 0.0112 m, i.e., it occupied 1% of the wick area. The wick properties and thickness were the same as for the experimental wick.

In Figure 9 the pressure field obtained from of one of the numerical calculations is shown. A hot spot of 15 times the normal heat flux has been imposed in position 2, Figure 8, where the normal heat flux is defined as the maximum uniform heat flux outside of the hot spot. The temperature for this case was 773 K, the working fluid potassium, the wick permeability 73×10^{-12} m^2 and the effective pore radius 50.8×10^{-6} m. The pressure isolines in the upper part of the wick are clearly affected by the presence of the hot spot. The distance between the 2000 Pa line and the 1500 Pa line is smallest at the horizontal position of the hot spot due to the high flow friction pressure drop around the hot spot. In general the vertical distance between the pressure isolines is increasing from the bottom of the wick and up; this is because the mass flow rate in the wick is decreasing upwards due to the evaporation. The minimum pressure ($P_{critical}$ = 1283 Pa, Equation (14)) is marked with a small circle on the top of the wick, i.e., in this case the dryout occurs at the top of the wick and not in the hot spot.

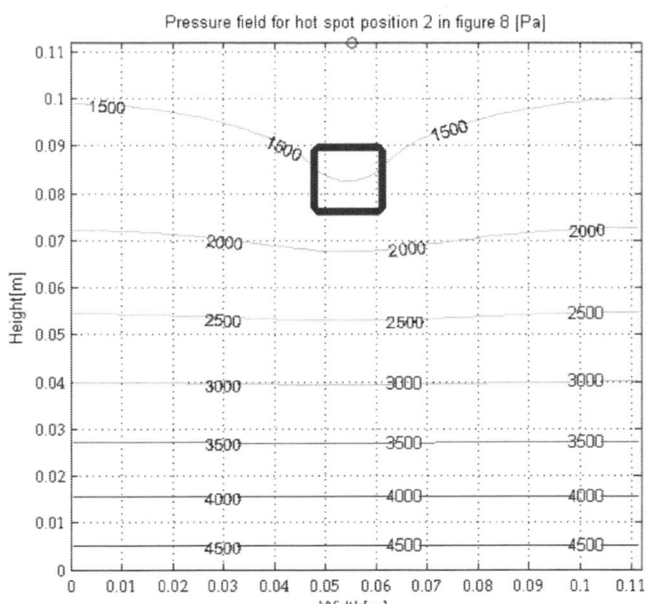

Figure 9. The pressure field for a 0.112 wide and 0.112 m high wick with a hot spot ($q_{hot\ spot}/q_{normal}$ = 15).

Figure 10 shows the maximum average heat flux, i.e., $q_{max,average} = (0.99 + 0.01 \cdot q_{hot\ spot}/q_{normal})q_{normal}$, for the case $q_{hot\ spot}/q_{normal} = 15$ as a function of the position of the hot spot. For a hot spot positioned below the vertical mid-position the average heat flux increases above the nominal value of 301.5 kW/m^2, whereas for positions above the middle the heat flux is lower than the nominal value.

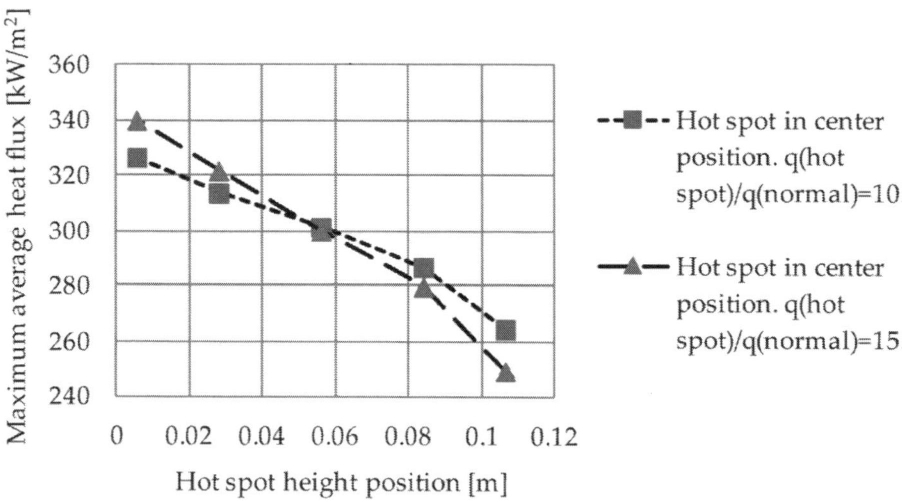

Figure 10. Maximum average heat flux as a function of hot spot vertical position.

Compared to the nominal case 301.5 kW/m^2 (without hot spot) a high flux hot spot above the mid-position will be more detrimental than a weaker hot spot. For a hot spot below the mid-position the reduced flow distance for the supply to the hot spot will be increasingly important with increasing hot spot heat flux.

In Figure 11 the wick maximum average heat flux, $q_{max,\ average}$, has been plotted for cases with hot spots $q_{hot\ spot}/q_{normal} = 10$ and $q_{hot\ spot}/q_{normal} = 15$ for different horizontal positions at the hot spot center vertical position (0.112–0.0112/2) = 0.1064 m.

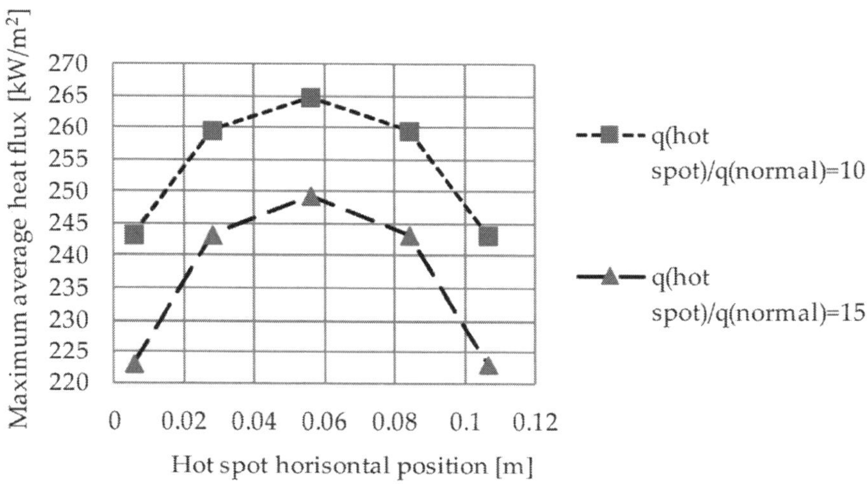

Figure 11. Maximum average heat flux for hot spots at different horizontal positions, all at top height level.

In Figure 11 the least detrimental hot spot positions are at the vertical centerline (i.e., for 0.056 m horizontal position), where the hot spots can receive liquid symmetrically from both sides. The flow resistance increases with distance from the centerline, and accordingly the maximum average heat flux decreases.

Figure 10 and Figure 11 show that the impact of the hot spot increases with its heat flux and that the vertical position of the hot spot is more important than the horizontal position for the analyzed cases.

Important observations from the analysis are:

- Positions 1 and 3 in Figure 8 are the two most detrimental hot spot areas for a vertical wick. The upper corner hot spot positions are critical because they are associated with the longest average flow distances for the liquid from the pool at the bottom of the wick and a higher local mass flow rate at the hot spot border. The hot spot positions 1 and 3 will lead to the lowest maximum average heat flux for the whole wick.

- Position 2 is less detrimental than position 1 and position 3 because a hot spot in position 2 will receive liquid from both sides, with a lower average flow resistance than the corner hot spots 1 and 3.
- A hot spot in position 7, 8, or 9 will lead to the highest maximum average heat fluxes because the average flow resistance is reduced compared to the uniform heat flux case. A hot spot in position 7, 8, or 9, which may not be sufficient to cause dryout there, can still cause dryout at the top of the wick.

A dryout situation caused by a hot spot will lead to a sharp increase in the temperature at the dryout position. Mechanical damage to the wick can occur due to the thermal expansion which accompanies the dryout. An undamaged wick will normally be rewetted at the dryout position if the heat flux at the hot spot is decreased below the dryout value.

One way to reduce the risk for hot spots is to increase the thickness of the wick, because the capacity of a wick is directly proportional to its thickness. Increased thickness will, however, increase the thermal resistance of the wick.

Reduced wick height will also increase the ability of the wick to handle hot spots, and lead to increased maximum average uniform heat flux.

The conclusions of the hot spot analysis are also considered valid for wicks exposed to non-uniform heat fluxes, in general.

Special design is required when the application requires cooling over a larger height than a single wick can handle. This may be a system consisting of several heat pipes, or a single heat pipe in which condensate is supplied to the wick at several vertical positions.

5. CONCLUSIONS

- From the analytical model:

  For the heat pipe at an operating temperature of 500 °C the thermal resistances related to the evaporation and condensation of the working fluid were small compared to the thermal resistances of the heat pipe walls.

- From the analytical model and the experimental results:

  A simplified thermal resistance spreadsheet model of the heat pipe produced operating temperatures which correlated within a ±10% error band within a wide range of operating parameters. The limited amount of experiments of this study indicated that the heat transfer capacity of the heat pipe utilizing potassium working fluid was higher than predicted from theory in the vapor static pressure-limited region, in the capillary pressure-limited region the capacity was in accordance with theory when the experimental uncertainty was taken into account. In general, the wick performance may be higher than predicted due to:

  A. The theory of Shirazy and Fréchette [17], which says that the effective pore radius is decreasing when the wick is exposed to a heat flux, i.e., use of an effective pore radius obtained from an experiment without heat flux will underestimate the wick capacity.

  B. The permeability of the wick is larger than measured in experiments with the model fluid. Point A can be excluded as explanation for the high wick capacity measured in the vapor static pressure limited region, only increased permeability can lead to increased performance in that part.

- From analysis of the thermal expansion by use of COMSOL Multiphysics®:

  The unilateral heat flux caused about 2 mm horizontal deformation (bending) of the experimental heat pipe. Asymmetrical thermal expansion caused by unilateral heat flux can cause thermal contact problems even for heat pipes of made of materials of relatively high thermal conductivity, like nickel, and can have created a high permeability channel in the wick in the current case.

- From the numerical 2D wick hot spot analysis:

  A rectangular, vertical wick is most vulnerable to hot spots located at the upper corners, which are associated with the highest flow resistances. Heat pipe technology is promising for an aluminum electrolysis cell wall cooling system due to high effective thermal conductivity and ability to handle hot spots.

ACKNOWLEDGMENTS

The authors would like to thank Elkem and the Research Council of Norway for funding this work.

AUTHOR CONTRIBUTIONS

Geir Hansen carried out the experimental work, data reduction, and prepared the first draft of the paper. The hot spot analysis was done by Kolbeinn Kristjansson. Geir Hansen and Erling Næss have equally contributed to the analysis and paper preparation.

CONFLICTS OF INTEREST

The authors declare no conflict of interest.

REFERENCES

1. Grjotheim, K.; Kvande, H. Introduction to Aluminium Electrolysis: Understanding the Hall-Héroult Process; Aluminium-Verlag: Düsseldorf, Germany, 1993.

2. Aune, J.A.; Johansen, K.; Nos, P.O. Electrolytic Cell for the Production of Aluminum and a Method for Maintaining a Crust on a Sidewall and for Recovering Electricity. U.S. Patent 6811677B2, 2 November 2004.

3. Faghri, A. Heat Pipe Science and Technology; Taylor & Francis: Washington, DC, USA, 1995.

4. Reay, D.A.; Kew, P.A.; McGlen, R.; Dunn, P.D. Heat Pipes: Theory, Design, and Applications; Butterworth-Heinemann, an Imprint of Elsevier: Oxford, UK, 2014.

5. Silverstein, C.C. Design and Technology of Heat Pipes for Cooling and Heat Exchange; Hemisphere Pub. Corp.: Washington, DC, USA, 1992.

6. Peterson, G.P. An Introduction to Heat Pipes: Modeling, Testing, and Applications; Wiley: New York, NY, USA, 1994.

7. Faghri, A. Review and advances in heat pipe science and technology. J. Heat Transf. **2012**, 134.

8. Reay, D.; Harvey, A. The role of heat pipes in intensified unit operations. Appl. Therm. Eng. **2013**, 57, 147–153.

9. Huang, X.; Franchi, G. Design and fabrication of hybrid bi-modal wick structure for heat pipe application. J. Porous Mater. **2008**, 15, 635–642.

10. Marcus, B.D.; Edwards, D.K. Graded Pore Size Heat Pipe Wick. U.S. Patent 4,170,262, 9 October 1979.

11. Chan, C.W.; Siqueiros, E.; Ling-Chin, J.; Royapoor, M.; Roskilly, A.P. Heat utilisation technologies: A critical review of heat pipes. Renew. Sustain. Energy Rev. **2015**, 50, 615–627.

12. Hansen, G.; Næss, E. Performance of compressed nickel foam wicks for flat vertical heat pipes. Appl. Therm. Eng. **2015**, 81, 359–367.

13. Queheillalt, D.T.; Carbajal, G.; Peterson, G.P.; Wadley, H.N.G. A multifunctional heat pipe sandwich panel structure. Int. J. Heat Mass Transf. **2008**, 51, 312–326.

14. Dullien, F.A. Porous Media: Fluid Transport and Pore Structure; Academic Press: San Diego, CA, USA, 2012.

15. International Organization for Standardization. Dense Shaped Refractory Products—Determination of Bulk Density, Apparent Porosity and True Porosity; ISO_5017; International Organization for Standardization: Geneva, Switzerland, 1998.

16. Holley, B.; Faghri, A. Permeability and effective pore radius measurements for heat pipe and fuel cell applications. Appl. Therm. Eng. **2006**, 26, 448–462.

17. Shirazy, M.R.S.; Fréchette, L.G. Effect of meniscus recession on the effective pore radius and capillary pumping of copper metal foams. J. Electron. Packag. **2014**, 136, 041003.

18. Dwyer, O.E. Boiling Liquid-Metal Heat Transfer, 1st ed.; American Nuclear Society: Hinsdale, IL, USA, 1976.

19. Rosenfeld, J.; Minnerly, K.; Dyson, C. Ten Year Operating Test Results and Post-Test Analysis of a 1/10 Segment Stirling Sodium Heat Pipe, Phase III; NASA/CR-2012-217430; National Aeronautics and Space Administration, Glenn Research Center: Cleveland, OH, USA, 2012.

20. Meisel, P.; Lippmann, W.; Hurtado, A. Ceramic high-temperature heat-pipes. In Proceedings of the 22nd International Conference on Nuclear Engineering, American Society of Mechanical Engineers, Prague, Czech Republic, 7–11 July 2014.

21. Hansen, G.; Næss, E.; Kristjansson, K. Sintered nickel powder wicks for flat vertical heat pipes. Energies **2015**, 8, 2337–2357.

22. Vargaftik, N.B.; Vinogradov, J.K.; Jargin, V.S. Handbook of Physical Properties of Liquids and Gases: Pure Substances and Mixtures; Begell House: New York, NY, USA, 1996.

23. Huang, X.Y.; Liu, C.Y. The pressure and velocity fields in the wick structure of a localized heated flat plate heat pipe. Int. J. Heat Mass Transf. **1996**, 39, 1325–1330.

24. Incropera, F.P.; DeWitt, D.P.; Bergman, T.L. Principles of Heat and Mass Transfer; Wiley: Singapore, Singapore, 2013.

25. Stephan, K.; Green, C.V. Heat Transfer in Condensation and Boiling; Springer: Berlin, Germany, 1992.

8

CHAPTER

THERMODYNAMIC ANALYSIS OF A WASTE HEAT DRIVEN VUILLEUMIER CYCLE HEAT PUMP

Yingbai Xie * and Xuejie Sun

Department of Power Engineering, North China Electric Power University, Baoding 07100, China

ABSTRACT

A Vuilleumier (VM) cycle heat pump is a closed gas cycle driven by heat energy. It has the highest performance among all known heat driven technologies. In this paper, two thermodynamic analyses, including energy and exergy analysis, are carried out to evaluate the application of a VM cycle heat pump for waste heat utilization. For a prototype VM cycle heat pump, equations for theoretical and actual cycles are established. Under the given conditions, the exergy efficiency for the theoretical cycle is 0.23 compared to 0.15 for the actual cycle. This is due to losses taking place in the actual cycle. Reheat losses and flow friction losses account for almost 83% of the total losses. Investigation of the effect of heat source temperature, cycle pressure and speed on the exergy efficiency indicate that the low temperature waste heat is a suitable heat source for a VM cycle heat pump. The selected cycle pressure should be higher than 100 MPa, and 200–300 rpm is the optimum speed.

Keywords: VM cycle heat pump; thermodynamic calculation; exergy efficiency analysis

1. INTRODUCTION

The energy crisis and environmental pollution are two of the world's major problems [1,2]. Utilization of waste heat is a better way to combat these problems, for this method not only supplies more energy, but also simultaneously reduces the environmental pollution [3]. The amount of waste heat is huge. For example, the energy consumption in China during 2013 was 3.75 billion tons coal equivalent, and the overall energy conversion efficiency is about 72.3% [4], which means the waste heat is over1 billion tons coal equivalent. However, how to access and use this waste energy is a tough problem.

Heat pump technology is recognized as the most efficient way to utilize the waste energy. The coefficient of performance (COP) of an electric or mechanical driven heat pump is always greater than 1. In some cases, such as when high temperature waste heat sources are available, heat driven heat pumps will be more preferable. Absorption, adsorption and steam ejection are the most common types [5]. In general, the COP of a double effect absorption heat pump would be greater than 1, up to 1.56. The COP of adsorption is typically under 0.7, and even lower for steam ejection technology. The Vuilleumier (VM) cycle is also a heat activated closed gas cycle. It can utilize thermal energy, such as solar energy, isotope radioactivity, waste heat, etc. Experimental COP can reach 1.61, which is the highest among heat driven heat pump technology [6].

The VM cycle was patented by Rudolph Vuilleumier in 1918 [7]. This cycle has many advantages, such as fewer moving parts, a closed cycle, making it easier to control the pollution, potential higher COP, lower noise, lower mechanical friction and usefulness for distributed energy systems with small power output [8–12]. This cycle can also be applied in the cryogenic field [13,14]. Unfortunately, disadvantages always exist along with advantages. The cylinder volumes are not isothermal but nearly adiabatic. Even if the heat exchangers and the regenerators are perfect, the COP will be lower than for the Carnot cycle because the mean temperature in the cylinder volumes differs from the temperature in the heat exchanger brine. For a cooling cycle based on the VM cycle, the temperature difference between the two cylinder volumes is often smaller than the change of temperature in the cylinder volumes themselves, and the COP for the cooling cycle is thus seriously reduced. Another disadvantage is the specific output is low. For a desirable heat output the machine must be large, and therefore rather expensive.

The first application of the VM cycle heat pump in the room temperature field was in 1989 by Carlsen. He designed a natural gas driven VM cycle heat pump with helium as refrigerant for residence heating. The heating capacity output is 7.5 kW, and the COP reaches 1.62 [15]. Later he developed a 20 kW gas fired VM cycle heat pump [16]. Kuehl transformed the crankshaft driven VM heat pump into the free piston driven type [17,18]. Lee analyzed the influence of structure on the performance of VM heat pump, and the results show that the structure of the regenerator has a significant effect on the performance [19]. Finkelstein [20], Sekiya [21,22] and Kawajiri[23] have presented analyses of system performance. Xie [24] carried out research on the influence of binary mixtures on the regenerator of a VM heat pump.

Up to now, no work has related the second law of thermodynamics to a VM cycle heat pump. To comprehensively understand the VM cycle heat pump, an exergy analysis is carried out along with an energy analysis in this paper to evaluate the application of a VM cycle heat pump in waste heat utilization.

2. VM CYCLE HEAT PUMP

2.1. Configuration

Figure 1 shows the construction schematic of a waste heat driven VM cycle heat pump. The basic VM cycle heat pump contains two cylinders, namely a cold cylinder and a hot cylinder. Each cylinder has a displacer which performs reciprocating motion inside the cylinder. A seal device is used between the displacer and the cylinders for anti-leakage purposes. The hot and cold displacers are driven by a crankcase with a connecting rod.

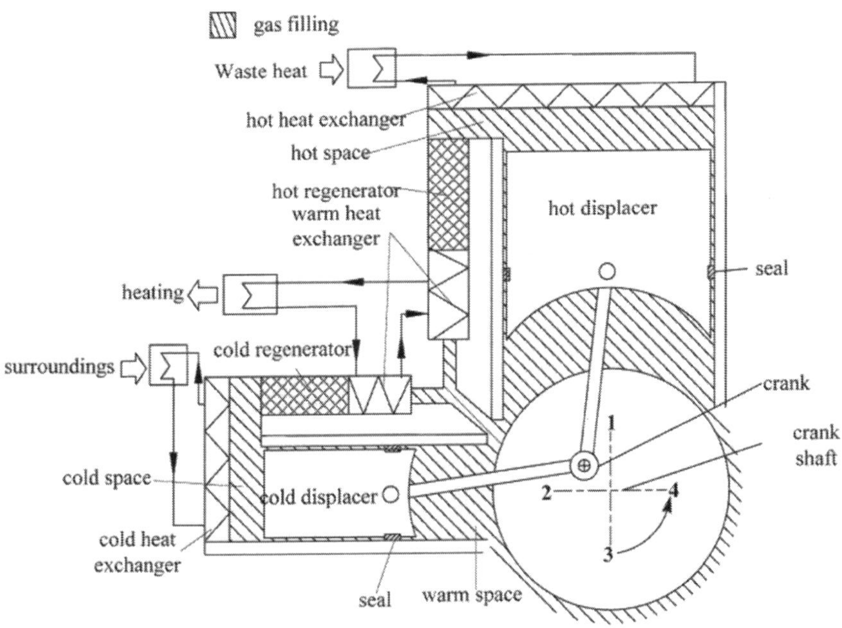

Figure 1. Schematic of a Vuilleumier cycle heat pump.

As shown in Figure 1, the space between the hot displacer and the top of the hot cylinder is called hot space, while the cold space is between the cold displacer and the top of the cold cylinder. The space between the cold displacer, hot displacer and the crankshaft is called warm space. A cold regenerator is placed between the cold space and warm space, while a hot regenerator is between the warm space and hot space. The regenerators work as a heat storage system when the working fluid flows between spaces.

The input high temperature waste heat is absorbed by the hot heat exchanger, mixed with low temperature heat from the surroundings which is absorbed by cold heat exchanger, providing heating from the warm heat exchanger for domestic purposes.

2.2. Working Process

The working process for a VM cycle heat pump is shown in Figure 2. The positions of the starting points are shown in Figure 1.

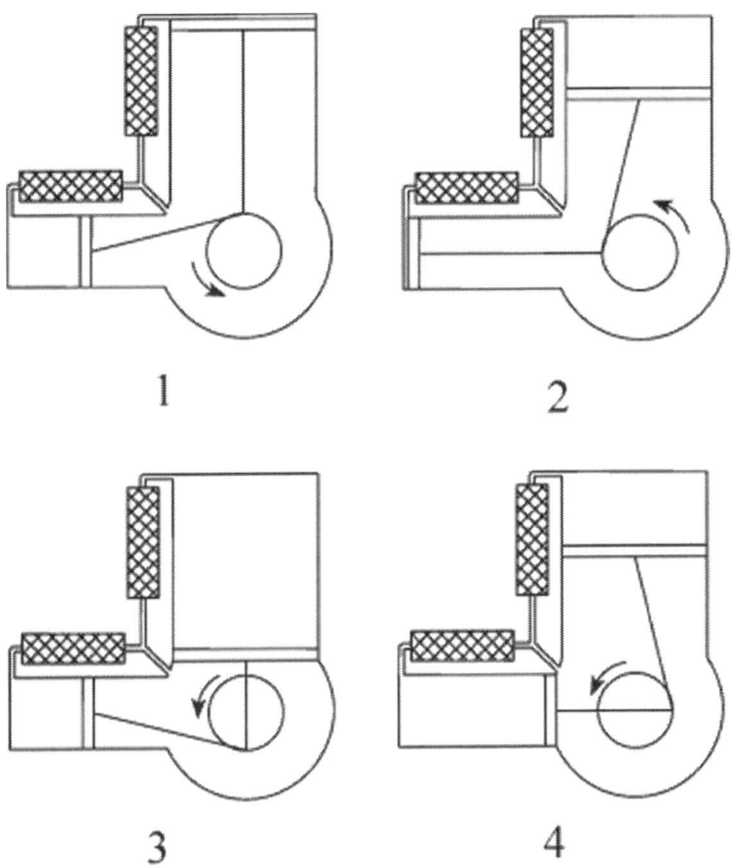

Figure 2. Working processes of a Vuilleumier cycle heat pump.

The cycle starts at point 1, and the whole process is illustrated as follows:

Process 1–2: the crankcase rotates counterclockwise, making the crankshaft push the hot displacer to the bottom of the hot space, while the cold displacer moves to the top of the cold space. During this process, the hot space volume increases, and the cold space volume decreases. As the hot displacer moves to the bottom, the working fluid from the warm space is heated to close to the hot space temperature (T_h) by filling materials when going through the hot regenerator, and then the working fluid goes into the hot space. The working fluid in the cold space is pushed by the cold displacer to the warm space, where it is heated to near the warm space temperature (T_a) by the filling material in the cold regenerator. Both the temperature and the pressure of the VM cycle heat pump increase rapidly. Output heat is supplied by the warm heat exchanger in the warm space.

Process 2–3: the crank moves from left to the bottom. The hot displacer still goes to the bottom of hot space. Meanwhile the cold displacer leaves the top to the bottom of cold space. The volumes of both are increasing. During this process, the working fluid in the warm space is cooled to nearly the cold space temperature (T_c) by the filling material in the cold regenerator, and then enters the cold space. Gas in the hot space absorbs heat from high temperature waste heat through the hot heat exchanger, while the cold space is absorbing heat from the low temperature surroundings using the cold heat exchanger.

Process 3–4: the crank rotates from bottom to right. The hot displacer leaves the bottom of the hot space to the top, and the cold displacer keeps moving to the bottom of the cold space. The hot space volume decreases, while the cold space volume keeps increasing. During this process, the working fluid in the hot space is pushed into the warm space by the hot displacer and cooled to nearly the warm space temperature (T_a) by the filling materials when crossing the hot regenerator. The gas temperature decreases, and the pressure also drops.

Process 4–1: the crank moves from right to the top. The hot displacer moves to the top of the hot space, and the cold displacer leaves the bottom to the top of cold space. The hot space volume decreases to nearly 0, and the cold space volume decreases continuously. During this process, gas in the hot space is cooled to nearly the warm space temperature (T_a) by the filling in the hot regenerator, then goes into warm space, with the pressure drop. A part of the gas in the cold space is heated to nearly the warm space temperature by the filling in the cold regenerator when it passes through the cold heat exchanger, and then returns to the warm space.

3. THERMODYNAMIC ANALYSIS

Simultaneous calculation and separate calculation are the two main methods to perform the thermodynamic analysis for a VM cycle heat pump [25]. In this paper, the separate calculation method based on an isothermal model is chosen. For the theoretical cycle, it assumes the following:

1) Processes occurring in all spaces are isothermal, i.e. the temperature of the compression or the expansion process is stable.
2) No losses occur in all the processes within the cycle, which means there is no heat transfer or flow resistance.
3) The working fluid is considered as ideal gas.

3.1. Working Fluid

The common working fluids adopted by the VM cycle heat pump are helium, hydrogen and nitrogen. As the VM cycle heat pump is a closed gas cycle, these working fluids are always in gas phase without any phase change. The properties of these working gases deviate from those of the ideal gas, so the different working gases adopted in a VM cycle heat pump have different system performance.

Generally hydrogen is the best working gas, but hydrogen is too active and it can easily explode when leaks occurs. Figure 3 shows that the differences in the thermodynamic performance of a VM cycle heat pump using hydrogen or helium as the working gas are small, and greater than when nitrogen is used as working gas. For both safety and performance concerns, helium is selected as the working gas for the VM cycle heat pump. All calculations carried out in this paper are based on helium as refrigerant. Some properties of helium are listed in Table 1.

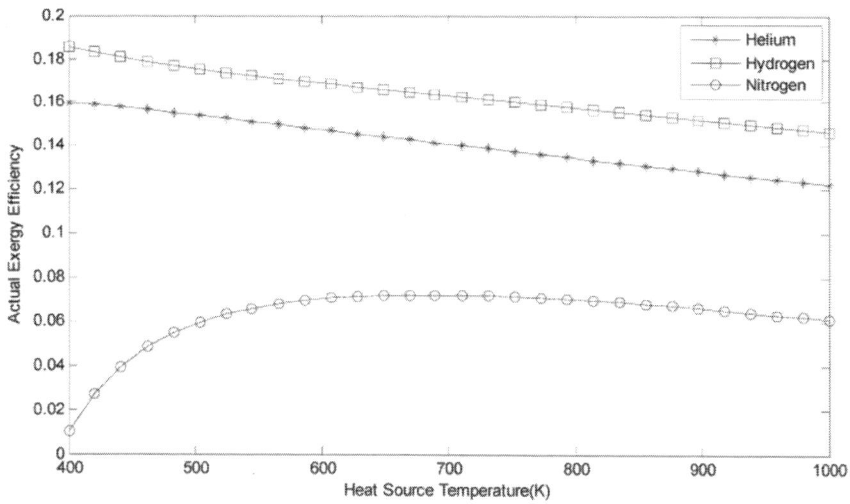

Figure 3. The influence of working fluids on exergy efficiency.

Table 1. Some properties of helium.

Parameter	Symbol	Value
Working fluid	*He*	—
Universal gas constant	R	8.314 J/(K·mol)
Specific heat	C_P	5.187×10^3J/(kg·K)
Thermal conductivity	λ	0.306W/(m·K)

3.2. Calculation Parameters

Table 2 gives the parameters needed for calculation, including constructional parameters and working conditions of the VM cycle heat pump.

Table 2. Primary calculation data.

Parameter	Symbolor Instruction	Value
Cylinder bore	D_{co}	0.0699 m
Stroke	Z	0.0312 m
Space phase angle	Φ	90°
Displacer length	L	0.04359 m
Displacer radial clearance	δ	0.00015 m
Regenerator diameter	D_R	0.0226 m
Regenerator length	L_R	0.0226 m
Regenerator material	Stainless steel wire	—
Hot space temperature	T_h	553 K
Warm space temperature	T_a	338 K
Cold space temperature	T_{co}	263 K
Surrounding temperature	T_0	273 K
Average temperature	$T_{av} = \frac{T_h + T_{co}}{2}$	408 K
Average pressure	P_{av}	10.0×10^6 Pa
Drive system	Single crank, Double-acting pistons drive	—
Rotational speed	n	600 rpm = 10 Hz
Volume ratio	ω	10

The heating output will be supplied as a radiator system or floor heating system. Radiator systems are more common, but the literature shows that floor heating systems are more comfortable. Floor heating systems have higher Predicted Mean Vote (PMV) and low vertical temperature differences. Floor heating systems also have energy saving advantages, as the system supply temperature is normally under 60 °C (typically it is 55 °C). The system return temperature is set at 35 °C.

3.3. Equations for the Energy and Exergy Analyses

All the equations used in this paper are referred to literature [8].

Heat absorbed by the cold space Q_{co}, referred to as the refrigerating capacity, can be defined as:

$$Q_{co} = P_{av} V_{co} \frac{\pi n \delta \sin\theta}{60\left(1+\sqrt{1-\delta^2}\right)} \tag{1}$$

where p_{av} is the average space pressure; V_{co} is the max volume of the cold space; n denotes the speed of the crank rotation; δ is the pressure parameter and θ is the phase angle.

Heat transferred from the waste heat in hot space Q_h is:

$$Q_h = Q_{co} \frac{\tau_{co} - 1}{1 - \tau_h} \tag{2}$$

where τ_{co} denotes the temperature ratio at the cold side, $\tau_{co} = T_a / T_c$; τ_h is the temperature ratio at the hot side, $\tau_h = T_a / T_h$.

Q_a is the heat output for heating purposes from the warm space:

$$Q_a = Q_{co} + Q_h \tag{3}$$

For an actual VM cycle heat pump, there exist lots of irreversible losses, such as reheat losses, flow friction losses, shuttle losses, pumping losses, heat conduction losses, leakage losses and so on. Reheat loss ΔQ_R represents the heat transfer losses in the heat exchanger of the regenerator caused by temperature differences, wall effects, temperature swings of materials, etc.:

$$\Delta Q_R = c_p M_0 (1 - \eta_R)(T_h - T_c) \tag{4}$$

where c_p is the specific heat at constant pressure; M_0 is the average mass flux through the regenerator; η_R is the regenerator efficiency, typically η_R =0.98.

Flow friction loss ΔQ_f refers to the pressure losses when the working gas passes through the regenerator, heat exchanger and ducts:

$$\Delta Q_f = \Delta P \frac{V_{co} n}{30} \tag{5}$$

where ΔP is the pressure drop.

In a VM cycle heat pump, displacers have the same axial temperature distribution as the cylinders. Due to the reciprocating motion of the displacers, there exists a temperature difference between the displacers and cylinders. This heat transfer loss is called shuttle loss ΔQ_{sh}:

$$\Delta Q_{sh} = F_s \frac{\lambda \pi D_{co} Z^2 (T_h - T_c)}{5.4 \sigma L} \tag{6}$$

where F_s is a correction factor, which generally has a value of 0.52–0.73; λ is the thermal conductivity, D_{co} is the diameter of the cold cylinder; Z is the displacer stroke; δ is the radial clearance between the displacer and cold cylinder; L is the displacer length.

Because of the temperature difference, the working gas entering each space will result in an extra heating load. This is called pumping loss ΔQ_p:

$$\Delta Q_p = \frac{2(\pi D_{co})^{0.6} L (P_{max} - P_{min})^{1.6} n^{1.6} c_p^{1.6} (T_h - T_c) \sigma^{2.6}}{1.5 Z_p (R')^{1.6} \lambda^{0.6} [(T_h + T_c)/2]^{1.6}} \tag{7}$$

where P_{max} is the maximum pressure during the cycle; P_{min} is the minimum pressure during the cycle; Z_p is the gas compressibility factor.

In an actual VM cycle heat pump, the length of the rod of displacers is short, resulting in heat conduction losses. The heat conduction loss ΔQ_d is the sum of the wall conduction loss ΔQ_{dw} and the filling material conduction loss ΔQ_{dm}:

$$\Delta Q_d = \Delta Q_{dw} + \Delta Q_{dm} \tag{8}$$

The wall conduction loss ΔQ_{dw} is:

$$\Delta Q_{dw} = \frac{A}{L}\left[\int_4^{T_a} \lambda(T)dT - \int_4^{T_c} \lambda(T)dT\right] \tag{9}$$

where A is the cross-sectional area.

Filling material conduction loss ΔQ_{dm} can be expressed as:

$$\Delta Q_{dm} = \frac{A}{L_R}\lambda_m (T_h - T_c) \tag{10}$$

There is also a temperature difference between the VM cycle heat pump and surroundings. This heat loss is called heat dissipation loss, ΔQ_l. We can take account for 5% of the Q_{co}, so the total losses of a VM cycle heat pump are:

$$\Delta Q = \Delta Q_R + \Delta Q_f + \Delta Q_{sh} + \Delta Q_p + \Delta Q_d + \Delta Q_l \tag{11}$$

The theoretical and practical aspects of thermodynamics are most relevant to energy and exergy analysis. Exergy analysis provides an alternative and illuminating means of assessing and comparing processes and systems rationally and meaningfully. Exergy analysis can assist in improving and optimizing designs.

The specific exergy flow of the working gas or water is evaluated as follows:

$$ex_{r,w} = (h - h_0) - T_0(s - s_0) \tag{12}$$

where h is the enthalpy; s is the entropy; and the subscript zero indicates properties at the reference (dead) state (i.e., at P_0 and T_0).

Here we suppose that the temperature T of the waste heat is constant, then the thermal exergy of waste heat at heat flow Q transferred to the VM cycle heat pump is:

$$Ex_Q = (1 - \frac{T_0}{T})Q \tag{13}$$

The exergy efficiency COP_{ex} [26] is generally expressed as the ratio of total exergy output to total exergy input and is written as follows:

$$COP_{ex} = \frac{Ex_{output}}{Ex_{input}} \tag{14}$$

where "output" refers to "net output" or "product" or "desired value", and "input" refers to "driving input" or "fuel". For a VM cycle heat pump, the output is the heating, with T_{out} at 55 °C and T_{in} at 35 °C, the input is the waste heat and/or the heat from the surrounding different than the dead state. Exergy efficiency for a VM cycle heat pump can then be written as follows:

$$COP_{ex} = \frac{E_{x,out} - E_{x,in}}{E_{x,h} + E_{x,c}} = \frac{Q_a\left(\left(1 - \frac{T_0}{T_{out}}\right) - \left(1 - \frac{T_0}{T_{in}}\right)\right)}{Q_h\left(1 - \frac{T_0}{T_h}\right) + Q_{co}\left(\frac{T_0}{T_c} - 1\right)} \tag{15}$$

3.4. Calculation Results

Table 3 shows the calculation results for the theoretical VM cycle heat pump. The assumptions of the theoretical VM cycle heat pump are as stated before, which include no irreversible losses. With the parameters in Table 3, the results show the theoretical exergy efficiency is 0.23.

Table 3. Theoretical calculation results.

Item	Symbol	Value
Stroke volume of cold space	V_{co}	119.73×10^{-6} m^3
Temperature ratio	τ	2.1
Pressure phase angle	θ	85.8°
Temperature ratio at cold side	τ_{co}	1.3
Temperature ratio at hot side	τ_h	0.6
Pressure factor	σ	0.2
Pressure ratio	P_{max}/P_{min}	1.5
Maximum pressure	P_{max}	12.4×10^6 Pa
Minimum pressure	P_{min}	8.1×10^6 Pa
Theoretical cooling of cold space	Q_{co}	4020.1 W
Theoretical heat consumption of hot space	Q_h	2948.7 W
Theoretical heat rejection of warm space	Q_a	6968.8 W
Theoretical exergy efficiency	$COP_{ex,i}$	0.23

Table 4 is the result of the actual process. The actual exergy efficiency is 0.15, much lower than the theoretical value of 0.23. The reason is the losses generated during the actual process. Among these losses, reheat loss and flow friction loss account for a large proportion of the total loss, almost 83%.

3.5. The Influence of Parameters on Exergy Efficiency

The theoretical and actual exergy efficiencies calculated above are based on the given parameters. Some parameters may affect the exergy efficiency greatly, while others only have a slight effect. The following is the analysis result of the effect of some parameters on the exergy efficiency for the calculated prototype VM cycle heat pump.

Table 4. Calculation results of the actual process.

Subject	Symbol	Value
Reheat loss	ΔQ_R	1317.9 W
Flow friction loss	ΔQ_f	1196.6 W
Shuttle loss	ΔQ_{sh}	322.4 W
Pumping loss	ΔQ_p	0.8 W
Wall heat conduction loss	ΔQ_{dw}	6.7 W
Filling heat conduction loss	ΔQ_{dm}	1.7 W
Heat conduction loss	ΔQ_d	8.4 W
Leakage loss	ΔQ_l	201.0 W
Total loss	ΔQ	3046.9 W

Table 4. *Cont.*

Subject	Symbol	Value
Actual cooling	$Q_{co,ac}$	1497.3 W
Actual heat rejection	$Q_{a,ac}$	4769.1 W
Actual heat consumption	$Q_{h,ac}$	3271.8 W
Actual exergy efficiency	$COP_{ex,e}$	0.15

Figure 4 shows the influence of heat source temperature on the exergy efficiency. As the temperature of the waste heat increases, the theoretical exergy efficiency decreases, while the actual first increases then decreases, reaching a maximum at the point of 370K. This means that we do not need to have much higher temperature waste heat since lower heat source temperature results in higher exergy efficiency. For most waste heat the temperature is under 600K, so a VM cycle heat pump can be used for waste heat utilization.

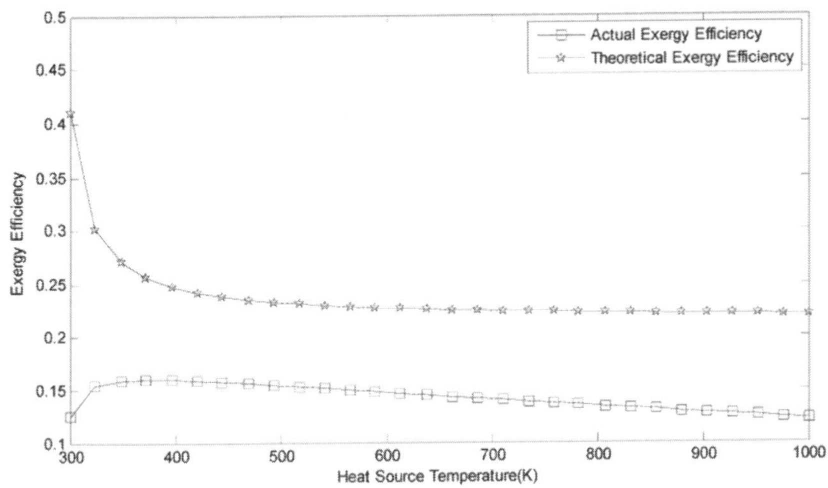

Figure 4. The influence of heat source temperature on exergy efficiency.

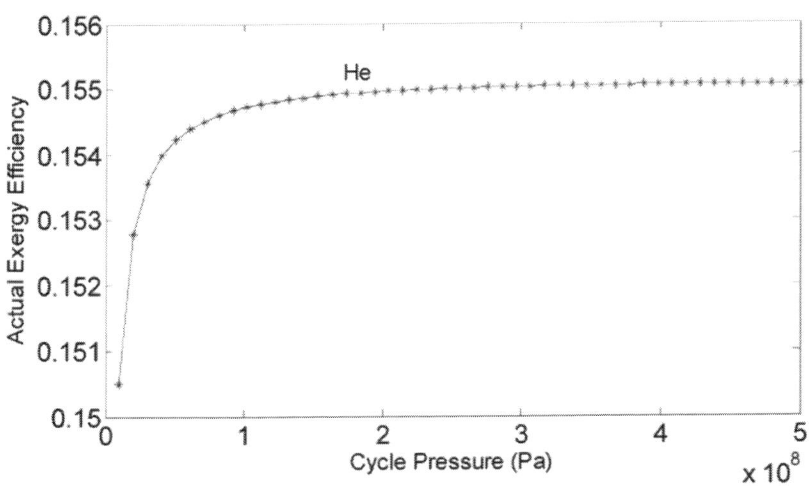

Figure 5. The influence of cycle pressure on the actual exergy efficiency.

The influence of cycle pressure on actual exergy efficiency is shown in Figure 5. As the pressure increases, the actual exergy efficiency also increases. After the cycle pressure reaches 100MPa, the exergy efficiency becomes stable. For most cases, the optimum pressure is hard to reach, so it can be treated as a reference to obtain higher performance.

Figure 6 presents the influence of speed and heat source temperature on actual exergy efficiency. A faster speed leads to lower exergy efficiency, while the amplitude of the exergy efficiency decrease increases.

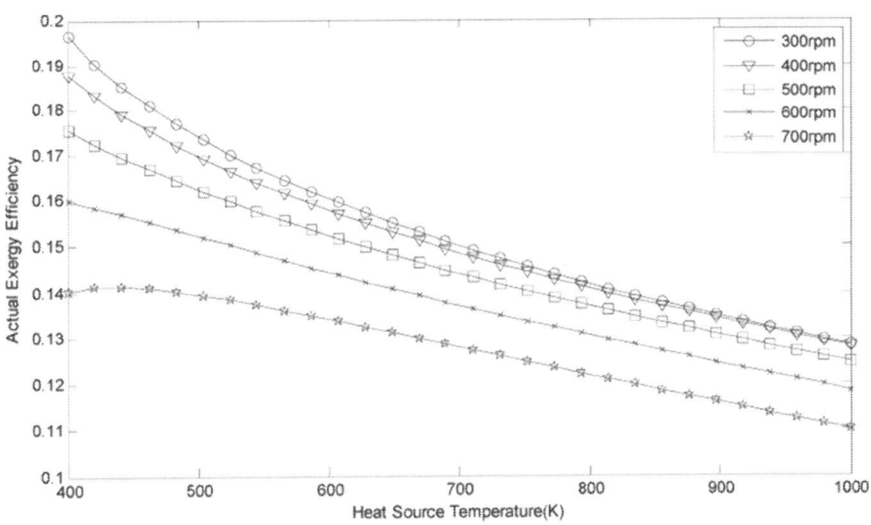

Figure 6. The influence of speed and heat source temperature on actual exergy efficiency.

4. CONCLUSIONS

In this paper, analyses of the theoretical and actual cycle of a VM heat pump are presented. Besides, the influence of some parameters on exergy efficiency is given. The conclusions are as follows:

1) According to the calculation results, the theoretical and actual exergy efficiency are 0.23 and 0.15, separately. Reheat losses and flow friction losses account for almost 83% of the total loss.

2) As the temperature of waste heat increases, the theoretical exergy efficiency decreases, while the actual one first increases then decreases, reaching a maximum at the point of 370K. Low temperature waste heat under 600K temperature is more suitable as the driving heat source of a VM cycle heat pump.

3) The cycle pressure and speed also have great influence on the exergy efficiency. The selected cycle pressure should be higher than 100MPa, and 200–300rpm is optimum for speed.

ACKNOWLEDGMENTS

The project is supported by the Natural Science Foundation of Hebei Province (No. E2014502085).

NOMENCLATURE

D Diameter (m)

Z Stroke (m)

Φ Volume phase angle (°)

L Length (m)

δ Clearance (m)

T Temperature (K)

P Pressure (Pa)

n	Rotational speed (rpm)
ω	Volume ratio
He	Helium
R	Universal gas constant ($JK^{-1}mol^{-1}$)
C_P	Specific heat ($Jkg^{-1}K^{-1}$)
V	Volume (m^3)
τ	Temperature ratio
Q	Heat quality (W)
θ	Pressure phase angle (°)
ψ	Porosity
D	Diameter (m)
A	Area (m^2)
a	Specific heat transfer area (m^2)
COP	Coefficient of performance
r	Hydraulic radius (m)
ρ	Density (kgm^{-3})
M	Mass flux (kgs^{-1})
ΔP	Pressure drop (Pa)
F_s	Correction factor
Δ	Heat loss (W)

Z_p Gas compressibility factor

COP_{ex} Exergy efficiency

Subscripts

co Cold space

h Hot space

a Warm space

R Regenerator

av Average

max Maximum

min Minimum

m Material

f Flow

sh Shuttle

p Pumping

i Ideal

e Effective

AUTHOR CONTRIBUTIONS

Yingbai Xie proposed and designed the research; Xuejie Sun performed the calculations. Both authors jointly worked on deriving the results and wrote the paper. Both authors have read and approved the final manuscript.

CONFLICTS OF INTEREST

The authors declare no conflict of interest.

REFERENCES

1. Grossman, P.Z.; Energy, Shocks. Crises and the Policy Process: A Review of Theory and Application. Energy Policy. **2015**, 8, 56–69.
2. Skouloudis, A.N.; Kassomenos, P. Combining Environment and Health Information Systems for the Assessment of Atmospheric Pollution on Human Health. Sci. Total Environ. **2014**, 488–489, 362–368.
3. Semkov, K.; Mooney, E.; Connolly, M.; Adley, C. Efficiency Improvement through Waste Heat Reduction. Appl. Therm. Eng. **2014**, 70, 716–722.
4. National Data, Available online: http://data.stats.gov.cn/workspace/index;jsessionid=0E028DD36CE5EEC4B0FDC6EB60547493?m=hgnd accessed on 13 March 2015.
5. Dong, L.; Huang, H.; Kobayashi, N. Development of an all-in-one Type Adsorption Heat Pump for Heating Application. Int. J. Chem. React. Eng. **2011**, 9, 18–24.
6. Xie, Y.B.; Deng, X.D.; Zhao, J.H. Fuzzy Comprehensive Evaluation of System Performance for Vuilleumier Cycle Heat Pump Driven by Solar, Geo-thermal and Biomass Energy. Appl. Power Energy Technol. **2014**, 8, 1080–1083.
7. Vuilleumier, R. Method and Apparatus for Including Heat Change. U.S. Patent 1, 275, 507 **1918**.
8. Bian, S. Cryogenic Refrigerator; China Machine Press: Beijing, China, 1990; pp. 56–57.

9. ASHRAE, Engine-driven heating and cooling equipment. In ASHRAE Systems and Equipment Handbook; Machinery Industry Press: Beijing, China, 1981; pp. 54–65.

10. Wu, Y. Refrigeration and Cryogenic Technology Principle; High Education Press: Beijing, China, 2007; pp. 26–33.

11. Li, X. Refrigeration Principle and Equipment; Machinery Industry Press: Beijing, China, 2006; p. 39.

12. Wang, R.; Ding, G.; Wu, J. Refrigeration Principles and Techniques; Science Press: Beijing, China, 2005; pp. 15–28.

13. Rule, T.T.; Qvale, E.B. Steady-Stale Operation of the Idealized Vuilleumier Refrigerator. Adv. Cryog. Eng. **1969**, 14, 343–352.

14. White, R. Vuilleumier Cycle Cryogenic Refrigeration; Technical Report AFFDL-TR-76-17; Air Force Wright Aeronautical Laboratories, Wright-Patterson Air Force Base: Dayton, OH, USA, 1976.

15. Carlsen, H. Development of a Gas Fired Vuilleumier Heat Pump for Residential Heating. Proceedings of the Intersociety Energy Conversion Engineering Conference, Washington, D.C., WA, USA, 6–11 August 1989; IEEE: New York, NY, USA, 1989; pp. 2257–2263.

16. Carlsen, H. Development of a New 20 kW Gas Fired Heat Pump Based on the Vuilleumier Cycle. Proceedings of the 25th Intersociety Energy Conversion Engineering Conference, Reno, NV, USA, 12–17 August 1990; IEEE: New York, NY, USA, 1990; pp. 233–238.

17. Kuhl, H.-D.; Schulz, S.; Thomas, B.; Carlsen, H. Conversion of a Crank-driven Vuilleumier Heat Pump into a Free Piston Machine. Proceedings of the Intersociety Energy Conversion Engineering Conference, Atlanta, GA, USA, 8–13 August 1993.

18. Pfeffer, T.; Kuehl, H.-D.; Schulz, S.; Walther, C. Development and Experimental Investigation of New Concepts for

Regenerators of Regenerative Gas Cycles-exemplary for Vuilleumier heat Pumps. Eng. Res. **2000**, 65, 257–272.

19. Lee, G.T.; Kang, B.H.; Yoo, H. Effects of Geometric Configuration on the Cooling Performance of a Vuilleumier Cycle Heat Pump. Proceedings of the Energy Conversion Engineering Conference, Washington, D.C., WA, USA, 11–16 August 1996; IEEE: New York, NY, USA, 1996; pp. 781–786.

20. Finkelstein, T. Isothermal Sinusoidal Analysis of Balanced Compound Vuilleumier Heat Pumps. Proceedings of the Intersociety Energy Conversion Engineering Conference, San Diego, CA, USA, 3–7 August 1992.

21. Sekiya, H.; Yamashita, I. Multisimulation Model for Stirling and Vuilleumier Cycle Machines. JSME Int. J. Ser. B Fluids Therm. Eng. **1993**, 36, 383–390.

22. Sekiya, H.; Kobayashi, K.; Fukuda, E. Numerical Analysis and Experimental Investigation of a Free Piston Vuilleumier Cycle Heat Pump. Proceedings of the 29th Intersociety Energy Conversion Engineering Conference, Washington, D.C., WA, USA, 7–11 August 1994.

23. Kawajiri, K.; Honda, T.; Sugimoto, T. Study of Free Piston Vuilleumier Heat Pump (Performance Characteristics of Prototype Machine at Forced Vibration). Trans. Jpn Soc. Mech. Eng. Ser. B **1996**, 62, 2430–2437.

24. Xie, Y.B.; Wang, S.H.; Li, B.; Liu, Y. Influences of Binary Mixtures on the Regenerator of VM Cycle Heat Pump. Appl. Mech. Mater. **2011**, 52–54, 249–254.

25. Rios, P.A.; Smith, J.L., Jr. An Analysis of Stirling Cycle Refrigerator. Adv. Cryog. Eng. **1969**, 14, 332–342.

26. Ozcan, H.; Dincer, I. Thermodynamic Analysis of a Solar Driven Tri-generation System for Building Applications. In Progress in Exergy, Energy, and the Environment; Springer International Publishing: New York, NY, USA, 2014; pp. 169–180.

9

CHAPTER

THERMAL EFFICIENCY COMPARISON OF BOREHOLE HEAT EXCHANGERS WITH DIFFERENT DRILLHOLE DIAMETERS

Jin Luo [1,*], Joachim Rohn [1], Manfred Bayer [2] and Anna Priess [1]

[1] *GeoZentrum Nordbayern, Friedrich Alexander University Erlangen Nuremberg, Bayern 91054, Germany*

[2] *TÜV Rheinland, LGA Bautechnik GmbH, Tillystr. 2, 90431 Nuremberg, Germany*

ABSTRACT

Thermal efficiency of borehole heat exchangers (BHE) is of crucial importance for the design and optimization of ground source heat pump (GSHP) system. This paper investigates thermal efficiency of a BHE with three drillhole diameters: 121 mm, 165 mm and 180 mm. The BHE was installed in a GSHP system of an office building located in Nuremberg, Germany. Thermal properties and hydraulic properties of the ground where the BHE was installed have been measured by thermal response tests as well as pumping tests. Furthermore, the evaluation of thermal performance is made possible by monitoring operation of the GSHP system. Using the recorded data, thermal exchange rates have been calculated and compared in a daily period as well as a seasonal period. The daily statistics indicate that the thermal exchange rate of the BHE increases with larger drillhole diameter. For the seasonal cooling performance, the amount of thermal exchange of BHE with 165 mm and 180 mm diameters was found to be 3.2% and 7.1% larger than that of the BHE with 121 mm diameter, respectively. These findings provide helpful suggestions for the design of future GSHP systems to achieve higher energy-efficiency.

Keywords: ground source heat pump system; borehole heat exchanger; drillhole diameter; thermal efficiency

1. INTRODUCTION

Recently, the increasing usage of fossil fuels has been inducing global warming as well as air pollution. Air conditioning of buildings accounts for a large part of the worldwide fossil energy consumption. In order to solve such pending problems, several new energy-efficient solutions have been developed in the last decades. Among all the technologies, ground source heat pump (GSHP) systems have been spotlighted as an efficient energy system due to their great energy savings and CO_2 emissions reduction potential. The thermal performance of GSHP system depends heavily on the heat transfer between a borehole heat exchanger (BHE) and its surrounding soil/rock. To ensure high efficiency of GSHP systems, thermal efficiency of the BHE is of essential importance [1,2,3].

Thermal performance of a GSHP system is commonly estimated by in-situ monitoring or by numerical modeling of the BHE. In the last decades, several studies [4,5,6,7,8,9] have been performed to assess the thermal performance of BHEs. These studies have discussed various factors affecting thermal performance such as geometric configuration of the BHE, ground conditions, etc. Pahud and Matthey [10] compared the thermal performance of a double U-tube BHE with different back fillings. Thermal performance was found to be enhanced with increasing thermal conductivity of the back fillings. Similarly, Borinaga-Treviño et al. [11] reported a study concerning the borehole thermal response and thermal resistance of BHE with four different grouting materials. The results indicated that the BHE with a cement-bentonite-graphite based grouting performs better than that of the BHE with cement-based grouting materials. The laboratory-measured grout thermal conductivity was found not to be suitable for borehole resistance estimation due to the increase of contact resistance between pipe and grout during the shrinkage of the cement based mix. Lee et al. [12] investigated the effective thermal conductivity in closed-loop vertical ground heat exchangers. Thermal efficiency was compared for six boreholes constructed with different grouting materials, different configurations as well as different additives. The testing results revealed that cement grouting has a higher effective thermal conductivity than that of bentonite grouting, and graphite performs better than silica. Besides, less thermal interference was found between the inlet and outlet pipe of a new 3-pipe type heat exchanger as compared to the conventional U-tube heat exchanger. Desmedt et al. [13] demonstrated an experimental study of thermal performance of borehole heat exchangers with different grouting materials. Thermal conductivity and borehole resistance were both measured in-situ by TRT tests. The borehole resistance for a standard coaxial tube with a cement-bentonite grouting was measured to be 0.344 K·m/W, whereas this value for a double U-tube was estimated to be 0.162 K·m/W which accounts for 48% of the coaxial tube.

Besides the grouting material, BHE configurations are also important for thermal performance. Zeng et al. [14] proposed an analytical solution considering the impacts of fluid axial convective heat transfer on borehole configurations. Different borehole configurations and flow circuit arrangements in regard to their thermal resistance have been assessed using this method. Results showed that the double U-tube performs better than a single U-tube. However, an opposite conclusion was obtained by Shu et al. [15] during the evaluation of thermal output for single and double U-tube with taken into account the economic performance. Therefore, both thermal and economic performance needs to be considered for an investment of a BHE. Choi et al. [16] investigated the effects of ground water flow on BHE arrays. Three types of BHE arrays have been examined using a two-dimensional numerical model. The results indicated that thermal performance of L-type and single line arrays are noticeably influenced by groundwater flow direction as well as flow rate.

The groundwater flow has also been reported to be an important factor in the thermal performance of BHEs. Wang et al. [17] demonstrated an experimental setup for analysis of the thermal performance of a BHE with consideration of groundwater flow. A model was developed to estimate the groundwater flow characteristics. The results showed that the groundwater flow has obvious impacts on the ground temperature profile in aquifers, which improves the thermal performance of the BHE. Zanchini et al. [18] examined the long-term performance of large borehole heat exchanger fields with unbalanced loads and groundwater flow. Peclet numbers in the range of 0–0.8 were considered in this study. The results show that the hourly peak loads was not affected by groundwater flow. However, the groundwater flow has an obviously influence on the long-term performance. On the other hand, Lazzari et al. [19] investigated the long-term performance of BHE fields with a negligible groundwater movement. The impacts of six time periodic heat loads on different field geometries were studied numerically. The findings indicated that no compensation between winter and summer loads is necessary for a single BHE. For a line of BHEs and for two staggered lines of staggered BHEs, at least a partial compensation of winter heating and summer cooling is needed.

To analyze in depth the thermal performance of a BHE, the short time response of a BHE as well as the thermal interaction between adjacent BHEs also need to be assessed. Lamarche and Beauchmp [20] developed an analytical model for the short time transient response of a BHE. Less than one hour heat transfer of BHE can be simulated using this model. The results from this analytical model follow very well the results from numerical models that take into account the thermal capacity of the borehole. Bernier et al. [21] studied the thermal interaction among boreholes in a geo-exchange system with vertical ground heat exchangers. The results indicated that long-term temperature changes have a significant impact on borehole field sizing and energy simulation results. In addition, the concept of temperature penalty which represents an effective increase of the undisturbed ground temperature was used for evaluation of borehole thermal interactions. A correlation was proposed to predict the temperature penalty and it provided results within an uncertainty of ±10% for most operation conditions.

However, the aforementioned studies were limited to small scales such as laboratory measurements and site tests. Some of them examined the thermal exchange in the BHE only based on numerical simulation. In this work, a more thorough study of thermal efficiency of BHE is conducted under practical GSHP system operation conditions. An analysis of the thermal efficiency of the BHE has been conducted on a GSHP system installed in an office building located in Nuremberg, Germany. The BHE with three drillhole diameters were installed in the GSHP system. The thermal performance is analyzed using the monitoring data accumulated under operation of the GSHP system. Thermal exchange between the BHE and its surrounding soil/rock is measured and analyzed. The remaining sections of this paper are organized as follows: Section 2 presents the ground properties measurement and the system configuration. Section 3 describes the measured data and analyzes the thermal performance of BHE under different diametric conditions. Finally, the article concludes with a summary of the results of this study in Section 4.

2. EXPERIMENTAL MEASUREMENTS

2.1. Ground Properties Measurement

2.1.1. Thermal Properties

Thermal properties of the ground where the BHE was installed have been measured by performing thermal response tests (TRT). For a TRT test, a defined heat load is input in the circulating heat carrier fluid and the resulting temperature changes are measured [22]. Figure 1 shows the recording of the TRT test performed for the drillhole of 121 mm diameter [23]. It is observed that the fluid temperature keeps increasing with testing time. Heat input remains almost constant during the whole testing period, except for a few minutes in the beginning of the tests. At the same time, the ambient temperature was observed to change abruptly, ranging from 3 °C to 30 °C during the 7-day testing period. These findings imply that the abrupt change in ambient temperature is an important factor affecting the testing results.

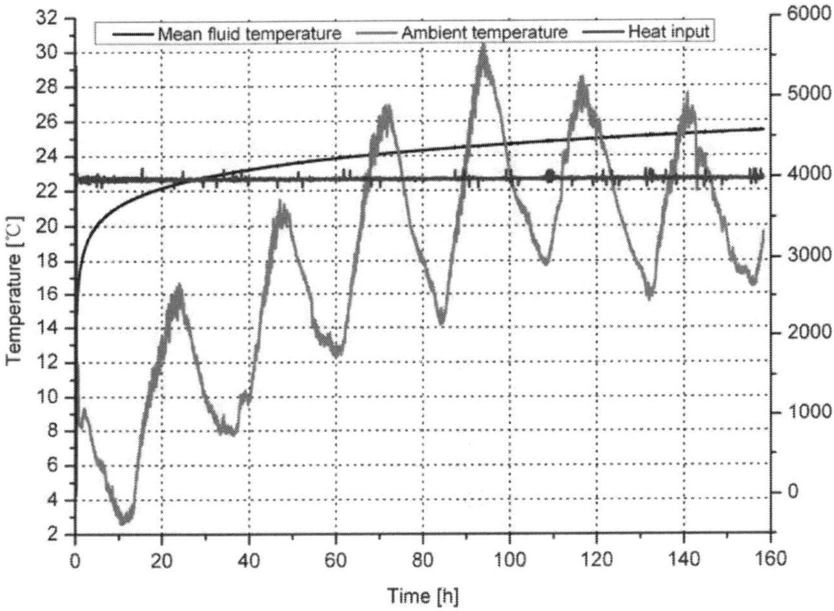

Figure 1. Recording of the TRT test performed in the drillhole of 121 mm diameter. The TRT tests have been performed separately at the three ensembles of the installed BHEs.

The TRT measurements were interpreted by following Hellström's method [24]. In this method, the heat transfer of a BHE in the ground is described by two parameters: effective thermal conductivity which describes the mean thermal conductivity of the surroundings for borehole, λ_{eff}; and borehole resistance which describes heat transfer between pipes and borehole wall, R_b. A steady-state heat transfer calculations gives a heat flow Q, between circulating fluid and borehole wall. Then, the heat transfer in a BHE can be derived as follows:

$$T_f = \frac{Q}{4\pi\lambda_{eff} H}\ln(t) + \frac{Q}{H}\left\{\frac{1}{4\pi\lambda_{eff}}\left[\ln\left(\frac{4a}{r_0^2}\right) - \gamma\right] + R_b\right\} + T_s \qquad (1)$$

where m is the inclination of the curve of temperature versus logarithmic time, to calculate effective thermal conductivity; λ_{eff}, the formula has to be transformed:

$$\lambda_{eff} = \frac{q}{4\pi Hm} \qquad (2)$$

In the first few hours, the temperature development is mainly controlled by the borehole filling and not by the surrounding soil or rock. Therefore, the minimum duration criterion should be noted. The minimum time for valid evaluation of the test is recommended as:

$$t > \frac{5r^2}{\alpha} \qquad (3)$$

The measured thermal conductivities for the three different drillhole diameters match nicely, which means average thermal conductivity along the borehole is around 2.5–2.6 W/m/K, as summarized in Table 1. Besides, drillhole diameter has an effect on the R_b, with bigger diameters yielding higher resistance. Similar findings were obtained by Smith and Perry [7] for tests conducted on BHEs using different grouts and varying pipe and drillhole diameters.

Table 1. Results obtained from the performed three TRT tests.

Drillhole diameter	λ (W/m/K)	R_b (m*K/W)
121 mm	2.5	0.093
165 mm	2.6	0.105
180 mm	2.6	0.110

2.1.2. Hydraulic Properties

To measure hydraulic properties, pumping tests have been performed. In order to perform the pumping tests, two investigation wells have been drilled with a depth of 80 m. One of the wells, the upstream, was named as groundwater measurement point one (GWM 1). The other one which located at the downstream was named as groundwater measurement point two (GWM 2).

According to the material types, the pumping tests have been performed in three different geological layers (Figure 2). Packers were used at each boundary between different layers to separate the tests. Thiem's method was used for interpretation of the horizontal pumping tests. In cases of that only one observation well at a distance, r, from the pumping well is available, the well discharge can be formulated as:

$$W = \frac{2\pi KD(s_w - s_1)}{\ln(r_1 / r_w)} \tag{4}$$

where, s_w and s_1 are the respective steady-state elevations of the water levels in the piezometers; r_1 is the distance between pumping well and observation well; r_w is the radius of the pumping well [24].

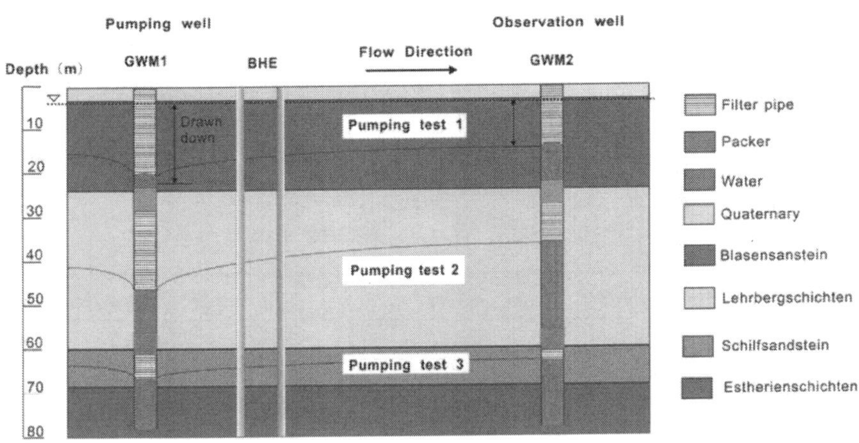

Figure 2. Schematic diagram of the pumping tests, the horizontal hydraulic properties of the geological layers have been investigated.

Hydraulic properties of the geological layers calculated using Thiem's method are listed in Table 2. The hydraulic conductivity of the "Blasensanstein" layer consisting of sandstone shows the biggest value of 1.71^{-5} m/s. That of the middle "Lehrbergschichten" layer which interbeds sandstone and claystone is determined to be 1.27^{-5} m/s. The hydraulic conductivity of the "Schilfsandstein" layer consisting of sandstone is measured to be 9.8^{-6} m/s. Furthermore, a permeability test was conducted to measure detailed material parameters of the layers. The hydraulic conductivity of the "Estherienschichten" layer that consists of claystone is measured to be 6.74^{-9} m/s. Therefore, this layer is treated as an aquiclude.

Table 2. The hydraulic properties estimated by pumping tests and permeability tests.

Geological layer	Material type	Depth (m)	Transmissivity (m^2/s)	Hydraulic conductivity (m/s)
Quaternary	sand	0–4	-	-
Blasensanstein	sandstone	4–25	3.4^{-4}	1.71^{-5}
Lehrbergschichten	Sandstone, claystone	26–55	3.8^{-4}	1.27^{-5}
Schilfsandstein	sandstone	56–62	6.88^{-5}	9.8^{-6}
Estherienschichten	claystone	62–80	-	6.74^{-9}

2.1.3. System Configuration

In order to investigate the thermal efficiency, in this study, the BHE was designed with three drillhole diameters using a commercial computer program Earth Energy Design (EED) [25]. By coupling into a GSHP system of an office building in Nuremberg city of Germany, the BHE was used for space heating as well as cooling. The office building, constructed in 2008, has three floors and one basement with a total area of 1530 m^2. The GSHP system was finished the installation works and started operation in 2009.

A schematic diagram of the installed GSHP system is depicted in Figure 3. Briefly, the GSHP system consists of a BHE, water-to-refrigerant heat exchanger and indoor units. The BHE was constructed with 18 boreholes that can be classified into blocks according to the drillhole diameter: block I of 121 mm, block II of 165 mm and block III of 180 mm, as shown in Figure 4. Geometrically, the BHE was a 2 × 9 array with a distance of 6 m between adjacent drillholes. In summer, the system operates in cooling mode and the heat obtained from the building is dissipated to ground. Whereas, in winter the GSHP system extracts heat from the ground to warm up the conditioned space using heat pump (SWP 75 I, Uponor GmbH, Hassfurt, Germany). The operation of this GSHP system either in heating mode or in cooling mode was controlled by a three-way valve.

The connection between the BHE and indoor units are the heat pump and the plate heat exchanger. The usage of the heat pump and the plate heat exchanger depends on the system operation mode. In winter, the system operates in heating mode and the heat pump is used for space heating. On the other hand, in summer, the system runs in cooling mode. A metal plate with a high thermal conductivity is used for transfer the energy from the fluid circulating in U-tube to the building. These components and a computer server were installed in the basement, covering an area of 380 m^2. The functional area of the building (offices, meeting rooms, corridors) are located in the three above-ground floors, covering a total air-conditioned zone of 1150 m^2. Coils were embedded into the floors and heat radiators were installed in the rooms to meet different temperature requirements (i.e. coils were designed to keep a constant temperature of 21 °C, heat radiators is 25 °C). The operation mode of the GSHP system depends on the actual energy demand of the building.

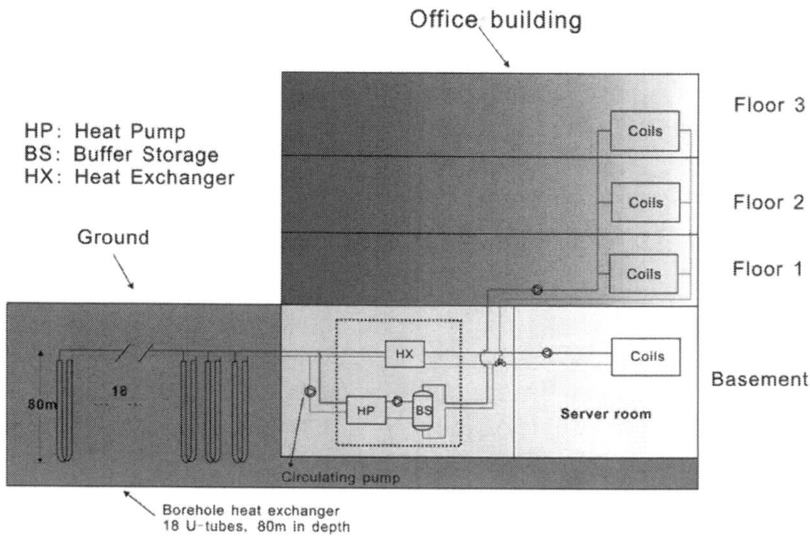

Figure 3. Schematic diagram of the GSHP system with vertical double U-tube BHEs. The BHEs of this GSHP system are in parallel connection.

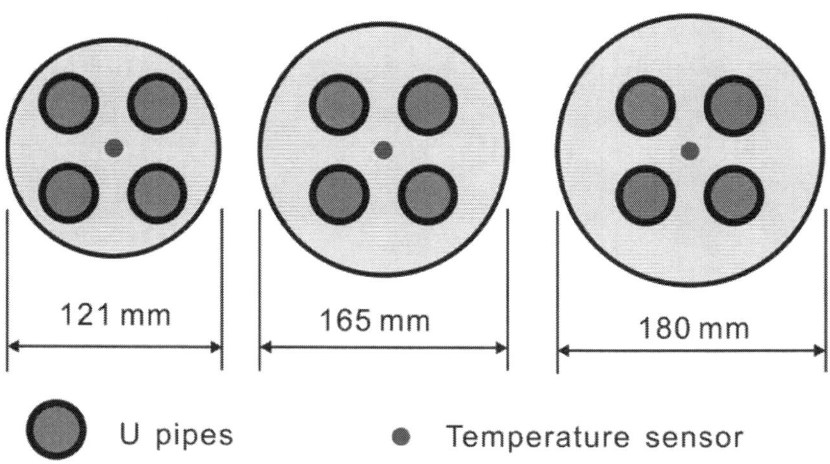

Figure 4. Schematic cross-section of the installed three-block BHEs. The temperature sensors were installed at the center of the BHEs and they are separately located at 18 m, 46.5 m and 77 m below the ground surface.

The sensors for measuring the subterranean borehole temperature are separately installed at 18 m, 46.5 m and 77 m below the ground surface. To keep the sensors at the same position, the cross-section center of the BHE, plastic holders were used. These holders with a hollow center were connected with four U-tube legs and the temperature sensors were installed in the hollow to measure the underground borehole temperature.

The specifications of the BHE components, containing configuration parameters and material properties are summarized in Table 3. It is observed that the geometric configurations of the three BHEs have the same inner configurations, but with different outer diameters. High density polyethylene double U-pipe was selected as the borehole heat exchanger. The space between the double U-pipe and the ground was grouted using enhanced cement with a thermal conductivity of 2.35 W/m/K. Glycol-based refrigerant media is circulated through the polyethylene pipe by water circulating pumps (TOP-S65/13, WILO AG, Dortmund, Germany). The properties of the grouting material are obtained from the products specifications.

Table 3. Geometric configurations and material properties for the BHE.

Components	Parameters	Specification
BHE configurations	Borehole	Borehole depth: 80 m
		Borehole distance: 6 m
		Drillhole diameter: 121 mm, 165 mm and 180 mm
	U-Pipe	Pipe type: Double-U
		Pipe shank spacing: 70 mm
		Pipe outer diameter: 32 mm
		Pipe wall thickness: 3 mm
BHE properties	Grouts	Grouting thermal conductivity: 2.35 W/m/K
		Volumetric water content: 31%
		Grouting density: 1.8 g/cm^3
	U-pipe	Pipe thermal conductivity: 0.42 W/m/K
Fluid properties	Heat carrier fluid	Fluid density: 1.11 g/cm^3
		Fluid volumetric thermal capacity: 3.8 MJ/m^3/K

2.2. Measurements Setup

Monitoring System

In order to investigate thermal efficiency of the BHE with different drillhole diameters, in this study, the measurements were conducted separately on each of the three BHEs. The temperature sensors (TST90, Endress + Hauser Messtechnik GmbH + Co. KG, Freiburg, Germany) were installed to measure fluid inlet and outlet temperature. At the same time, flow meters (Promag 53P40, Endress + Hauser Messtechnik GmbH + Co. KG) were installed to monitor flow rate of the fluid circulating through the U-pipes. Figure 5 shows schematic diagram of the measurement setup. It can be seen that the pipes for each six BHEs that have the same drillhole diameter were assembled into a chamber. The sensors were installed inside the chambers to monitor the fluid. Consequently, the fluid temperature and flow rate had been monitored separately for each drillhole diameter.

The specifications of these sensors, together with their models and measuring accuracy are listed in Table 4. The operating conditions of the GSHP system can then be monitored using these sensors. More specifically, the installation position of the sensors is also mentioned in this Table. The detailed installations of these sensors are depicted in Figure 4, Figure 5 and Figure 6.

Figure 6 shows a snapshot of the measurement setup on site with a view of internal settings of the plastic chamber. Two manifolds were installed for each plastic chamber, the lower one was for the fluid circulating to ground and the upper one was for the fluid circulating to the building. Pipes of the BHE were separately assembled for fluid inlet and outlet, which indicate that each six BHEs are parallelly connected. Therefore, the flow rate was evenly distributed due to the same pressure distribution of each pipe. In addition, outdoor temperature was also monitored using the sensor (TST434, Endress+Hauser Messtechnik GmbH + Co. KG). Then, thermal performance under different environmental conditions was analyzed. All the information regarding temperature, flow rate and the energy budget were continuously recorded by a data acquisition system. These parameters, together with the date and the time of the day, were also stored every 10 min.

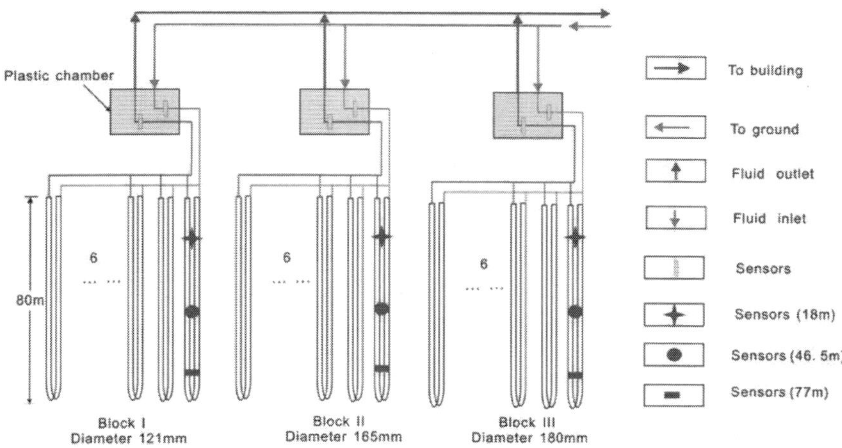

Figure 5. Schematic diagram of the measurement setup. The temperature sensors were separately installed at each block to measure the U-tube fluid in-outlet. There are twelve sensors, containing six temperature sensors and six flow meters. In addition, the temperature sensors were also installed inside the BHEs with different depths: 18 m, 46.5 m and 77 m. We select one of the sensors which have the same position in each block to measure the subterranean borehole temperature.

Table 4. The installed temperature sensors and flow meters [26].

Monitoring parameter	Components	Specification
Subsurface borehole temperature	PT-100 temperature sensors	Manufacture: Endress + Hauser Messtechnik GmbH + Co. KG Model: TST434 Measure range: −50–100 °C Uncertainty: ±0.05 °C Located at depth: 18 m, 46.5 m and 77 m
In-outlet fluid temperature	PT-100 temperature sensors	Manufacture: Endress + Hauser Messtechnik GmbH + Co. KG Model: TST90 Measure range: −50–200 °C Uncertainty: ±0.05 °C Located at fluid inlet/outlet of U-tube
Flow rate	Flow meter	Manufacture: Endress + Hauser Messtechnik GmbH + Co. KG Model: Promag 53P40, DN40 1 1/2" Assists energy: voltage 85–260 V Measure range: 0 dm^3/min–200 dm^3/min Uncertainty: ±0.2% Located at fluid inlet/outlet of U-tube
Energy budget	Data logger	Manufacture: Endress + Hauser Messtechnik GmbH + Co. KG Model: Memograph M RSG40 The results are stored every 500 microseconds

Figure 6. A snapshot of the measurements onsite. Two manifolds were installed, the lower one was for the fluid circulating to ground and the upper one was for the fluid circulating to the building. The marked 1, 2, 3, 4, 5 and 6 are for fluid circulate to the building; I, II, III, IV, V and VI are fluid circulate to the ground.

Based on the measurements, the energy exchanged between the BHE and the surrounding soil/rock at two successive entries of data logging can be calculated as:

$$E = q\rho c_m (T_o - T_i) \quad (5)$$

where E is the thermal exchange rate (kW); q is the flow rate (m^3/h); ρ is the fluid density (g/cm^3); T_i is the fluid inlet temperature (°C) and T_o is the outlet temperature (°C) [26]. The instantaneous (500 microsecond interval) heat supplied to the building was calculated by Equation (1) and the mean values with every 10min interval were stored. These values of thermal load can be positive or negative. Conventionally, the heat supplied to the building is considered to be positive.

3. RESULTS AND DISCUSSION

3.1. Typical Daily Operation

We selected August 07, which was a typical summer day, to investigate operation of the GSHP system in cooling mode. Figure 7 shows the ambient temperature, fluid temperature at inlet/outlet of the ground and the corresponding flow rate in that day. The fluid inlet temperature was higher than the fluid outlet temperature, which indicates that the ground was heated by the energy removed from the building. The flow rate of the circulating fluid maintains zero when the ambient temperature dropped to 20 °C. During this period, between 01:00 AM and 09:00 AM, the GSHP system stopped operation and the fluid temperatures (inlet and outlet) were found to increase obviously. This can be attributed to the heat conduction between the circulating fluid and its surroundings at the position of BHE top where with a depth of 1.0 m. At such a depth, the soil was heated by the solar radiation during the daytime and this resulted in a high ground temperature. Therefore, the fluid inlet/outlet temperature measured at the top of the BHE was observed to increase. On the other hand, when the GSHP system was operating, the fluid temperature was almost constant, regardless of any abrupt changes in ambient temperature. The fluid inlet-outlet temperature difference was maintained at 1 °C. This fact implies that the GSHP system has a stable cooling performance in summer.

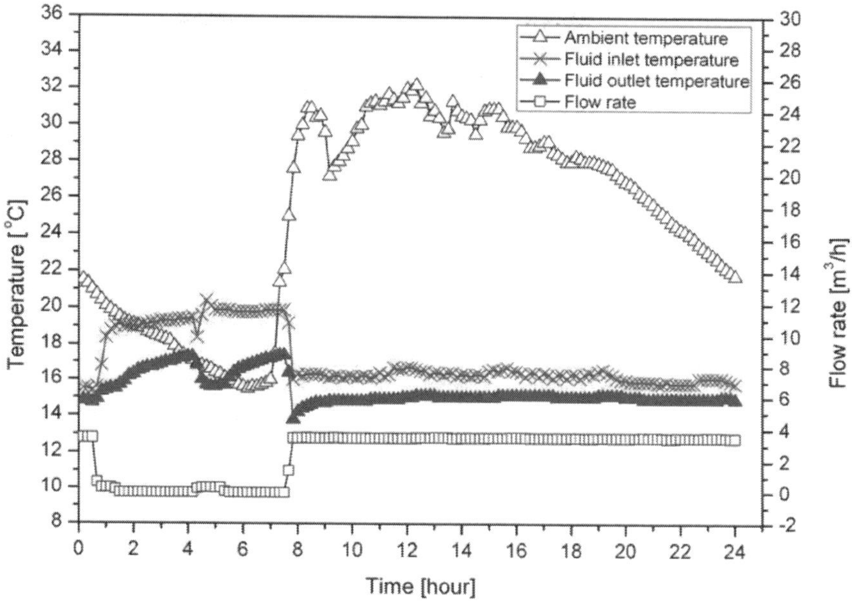

Figure 7. Recording of the GSHP system operation in a typical summer day (August 07, 2009).

Figure 8 represents the operation conditions of the GSHP system in heating mode. The measurements were conducted on December 27, a typical winter day with the ambient temperature varying from −4 °C to 4 °C. The fluid inlet temperature is lower than the fluid outlet temperature, indicating that the ground was used as a "heat source". It is observed that the flow rate of the heat carrier fluid fluctuates frequently between zero and 4 m³/h. At the same time, the fluid temperature fluctuates as well. Those fluctuations can be easily explained by the operation of the heat pump. In winter, the heat pump operated to extract energy from the heat carrier fluid and the fluid temperature then reduced sharply. Due to the higher ground temperature the fluid temperature can be recovered rapidly when the heat pump stopped operation. Consequently, the fluid temperature and flow rate fluctuated as observed. This finding indicates that electricity consumption of the building can also be fluctuated due to the aforementioned operation of the heat pump.

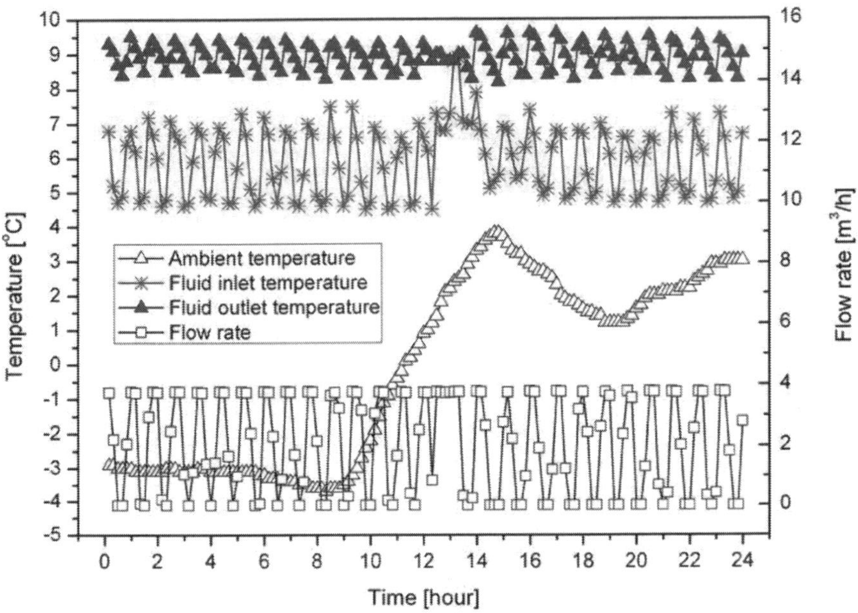

Figure 8. Recording of the GSHP system operation in a typical winter day (December 27, 2009).

3.2. Comparison of Experimental Condition

In order to compare the experimental conditions, the fluid inlet temperature of the three blocks BHEs was investigated. Figure 9 represents the hourly fluid inlet temperature measured on August 07, a typical summer day. Minor difference of fluid temperature was found in the three investigated BHEs, except for the period between 01:00 AM and 08:00 AM. During this period, as shown in Figure 7, the GSHP system was shut down. The noticeable differentiation in fluid temperature was therefore caused by the heat conduction between soil and static fluid in the pipe as aforementioned in typical daily operation. To evaluate the deviations of the three blocks, the fluid inlet temperature was compared during the system operation. The fluid temperature differences of the investigated three blocks are quite small, within 0.1 °C. Considering that the measurement uncertainties of the temperature are ±0.05 °C, such a small difference can then be considered negligible.

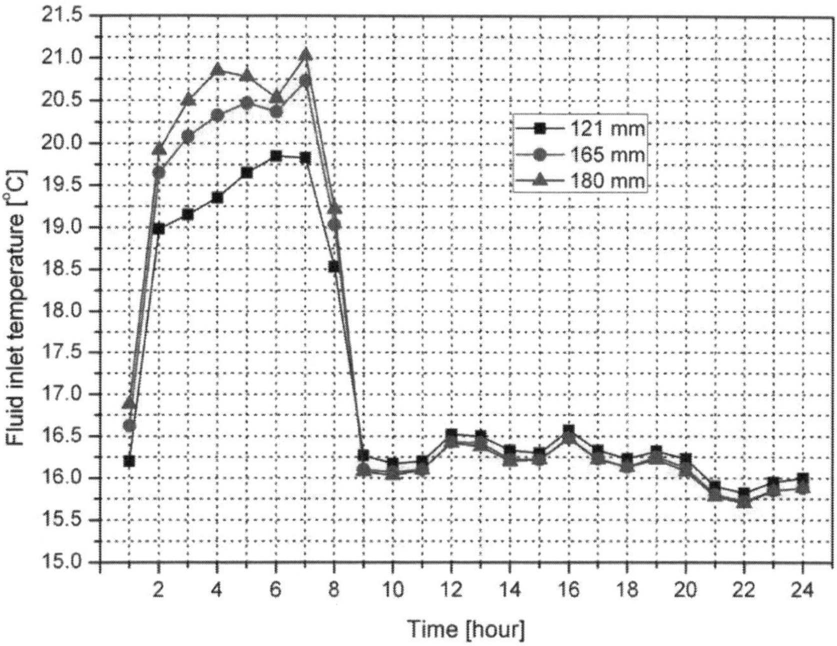

Figure 9. Hourly fluid inlet temperature distribution of the three investigated three blocks BHEs on August 07, 2009.

On the other hand, the fluid inlet temperature was also investigated on a typical winter day. Figure 10 shows the hourly fluid inlet temperature profile on December 27, 2009. The fluid inlet temperature was observed to drop severely at the BHE of 121 mm diameter when compared to the other two blocks (block II and block III) which have a negligible difference in the fluid inlet temperature. As shown in Figure 5, the BHE of 121 mm diameter have the largest distance to the building, whereas, block II and block III have similar distances that are both smaller than that of the 121 mm diameter BHE. Considering the intermittent operation of GSHP system and the ground cooled by the chilly weather in winter, as aforementioned in Figure 8, the serious temperature drop can be well explained by heat dispersion to the ground due to the lower temperature of the surrounding soil. This finding indicates a large amount of energy loss in the horizontal pipe during winter day system operation. To achieve higher energy-efficiency, the energy loss in the horizontal pipe needs to be seriously considered in future GSHP system design and implementation in cold regions. Meanwhile, the unevenly distributed fluid inlet temperature in three BHEs ensembles suggests also that comparison of thermal performance using the data collected from winter days may deviate. Therefore, the thermal efficiency comparison, in this paper, is conducted using only the data collected during the GSHP system in cooling mode to ensure higher accuracy.

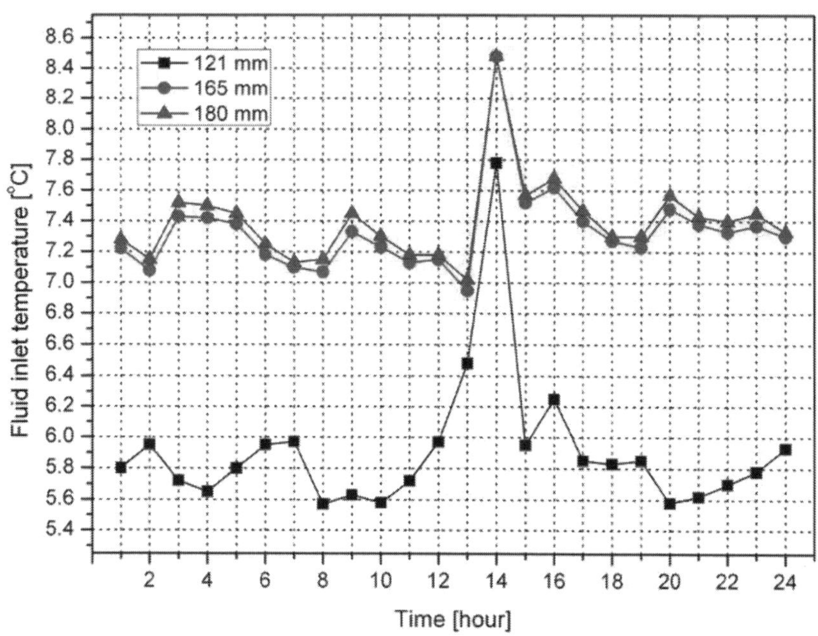

Figure 10. Hourly fluid-inlet temperature distribution of the three investigated BHEs at December 27, 2009.

3.3. Comparison of Thermal Efficiency

To compare thermal efficiency more accurately, this study is solely based on data of the GSHP system operating in heating mode in 2009, the beginning of operation of the GSHP system. The data collection started from July till the end of that year. The thermal performance is estimated considering the amount of thermal exchange which can be calculated with Equation (5). Based on the thermal exchange, the thermal efficiency can then be compared as follows:

$$R_i = \frac{E_i - E_{121}}{E_{121}} \times 100\% \tag{6}$$

where R is the thermal comparison ratio (%); E is the amount of thermal exchange (kWh); and the subscript i is the object that need to be compared with the BHE with a diameter of 121 mm.

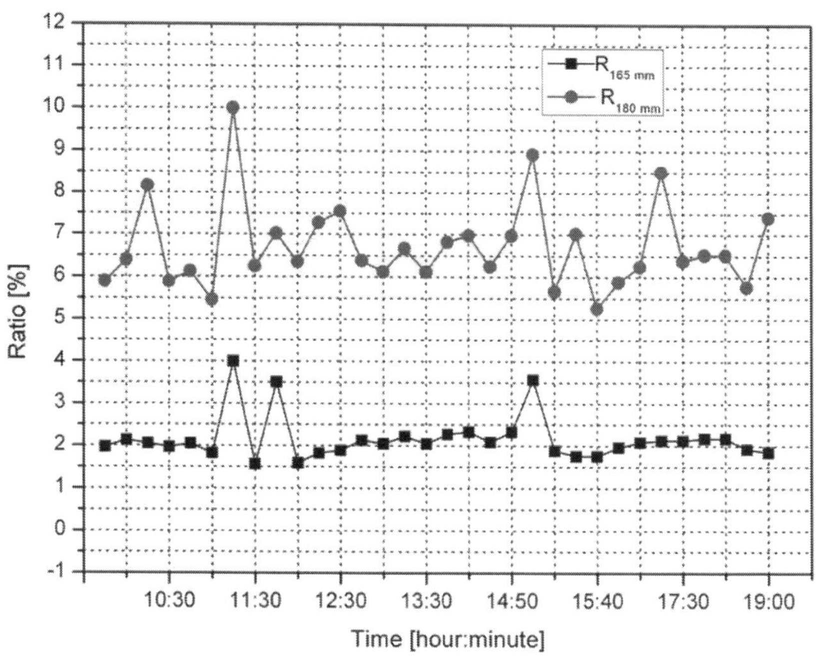

Figure 11. Thermal efficiency comparison of the BHE with three different drillholes on August 07, 2009. The thermal comparison ratio is presented for block II and block III as compared to block I.

Figure 11 represents the comparison results obtained using the data measured on August 07, 2009. These measurements were only conducted while the GSHP system was in operation from 10:00 AM to 05:00 PM on that day. The comparison of the thermal efficiency shows that the BHE of 180 mm diameter has the best thermal performance. On average, the daily amount of thermal exchange in the BHE of 180 mm diameter was calculated to be 6.7% larger than that the BHE of 121 mm diameter. The maximum difference between the 121 mm and 180 mm diameters BHEs was 10% in that day. For the comparison of BHE of 165 mm diameter and 121 mm diameter, the daily average ratio drops down to 2.16%, accounting for 54% of the daily maximum value (4.0%). These results indicate that increase the drillhole diameter will slightly improve the thermal performance of BHE despite the fact that the thermal conductivity of grouting material (2.35 W/m/K) is smaller than that of the ground (2.5 W/m/K).

Another fact regarding the GSHP system operation is revealed in Figure 12. As it can be seen, monthly thermal loads for the BHEs against ambient temperature are presented separately. Thermal loads of the three blocks of BHEs show an obvious increasing trend. Like the daily investigation, thermal performance of the BHE increases for larger drillhole diameter. At the same time, the thermal load of the BHEs is observed to strongly correlate with the ambient temperature. With higher ambient temperature, the cooling load increases as well.

Figure 13 shows the investigated thermal comparison ratio and the subterranean borehole temperature. The subterranean borehole temperature, in this work, was measured in the middle of the BHE, as shown in Figure 4. The mean value of the temperature measured at three depths is used due to the fact there is a negligible difference among these three depths. It is noticed that the difference of cooling performance for the BHE with different diameters is negligible when the subterranean borehole temperature reaches up to 12.5 °C.

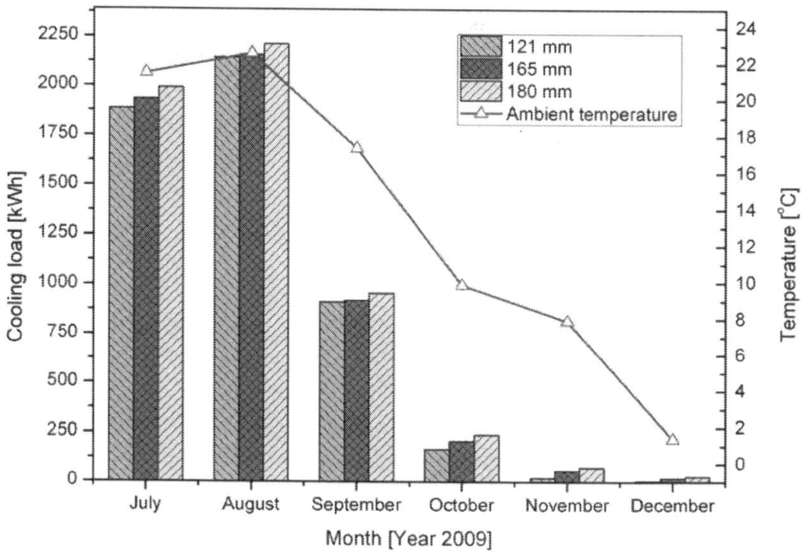

Figure 12. Monthly cooling load between July, 2009 and December, 2009.

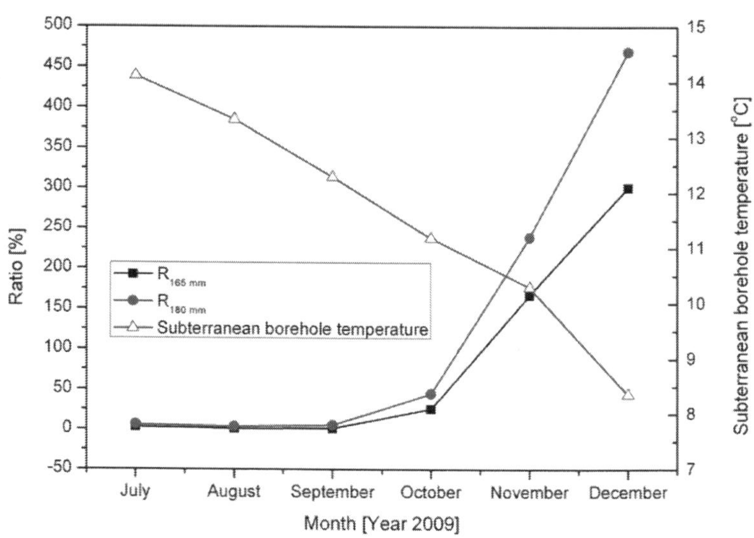

Figure 13. Thermal efficiency comparison of the BHE under different subterranean borehole temperature.

When the subterranean borehole temperature drops down to 11 °C, the difference becomes substantial. Even more, this difference will be several times larger when the subterranean borehole temperatures are lower than 10 °C. These findings suggest that thermal efficiency of the BHE with different diameters depends heavily on the subterranean borehole temperature. With a lower subterranean borehole temperature yields a higher cooling performance of the BHE with a bigger diameter.

In the long run, the thermal efficiency of the BHEs was estimated by calculating the seasonal cooling performance. Figure 14 represents the seasonal amount of thermal exchange between the BHE and its surroundings. The results show an ascending trend of the amounts of heat with increased drillhole diameter. Comparing the three BHE ensembles, the amount of seasonal thermal exchange for the BHE of 165 mm and 180 mm diameter is 3.2% and 7.1% larger than that of 121 mm diameter, respectively.

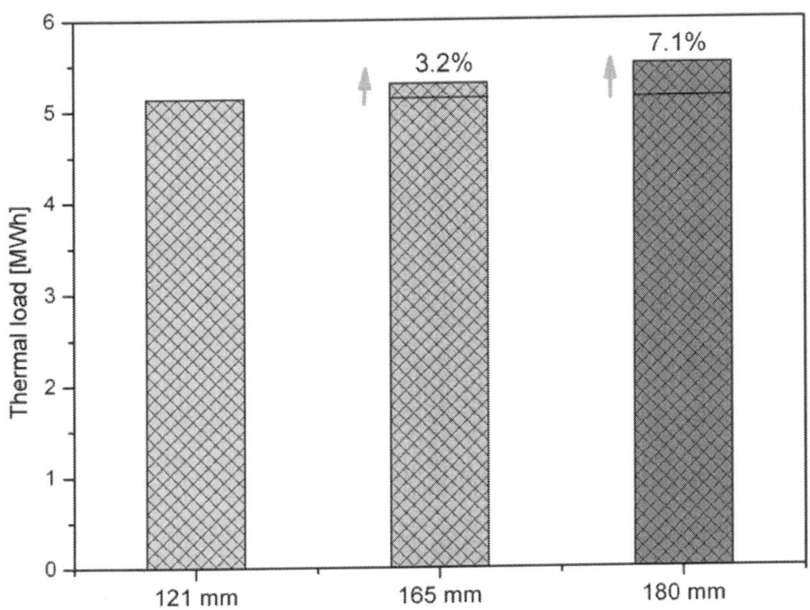

Figure 14. Comparison of seasonal thermal efficiency of BHE with different drillhole diameters.

4. CONCLUSIONS

In this paper, the thermal performance of a borehole heat exchanger (BHE) was investigated for three different drillhole diameters. This study was conducted by monitoring the operation of the GSHP system in 2009. The daily GSHP system operation was discussed separately for heating mode and cooling mode. In addition, the impacts of ambient temperature and subterranean borehole temperature on thermal performance were also analyzed. Finally, the seasonal thermal efficiency of the BHE was compared and discussed. Major conclusions are as follows:

- Thermal properties of the ground where the BHE was installed were measured by thermal response (TRT) tests. The effective thermal conductivity of the ground with a depth of 80 m was measured to be 2.5–2.6 W/m/K. This value is bigger than that of the grouting material with a thermal conductivity of 2.35 W/m/K. During performing the TRT tests, the ambient temperature was found to be an important factor to influence the testing results. Therefore, the connection pipes for a TRT test need to be carefully insulated. In addition, the hydraulic properties of the geological layers were measured by pumping tests. Based on the measurements, the geological layers were classified into different hydraulic units;

- The GSHP system was investigated for both a typical winter day and summer day. On a typical summer day, the GSHP system operates in cooling mode. On that day, the GSHP system delivers a stable cooling performance due to the constant fluid temperature and flow rate. On a typical winter day the heat pump operates to extract the energy from the heat carrier fluid, which abruptly changes the fluid temperature. Conclusively, the fluid temperature and flow rate fluctuate with time;

- In order to ensure the same experimental conditions, the fluid inlet temperature was investigated. Only negligible differences of the fluid inlet temperature were found in the three BHE blocks. However, an obvious fluid temperature drop was observed when the fluid circulates in the horizontal pipe in winter. This fluid temperature drop needs to be seriously considered in future

GSHP system design and implementation in cold regions. As a result, the comparison of thermal efficiency was conducted using only the data collected during summer;

- The comparison of the daily thermal efficiency suggested that the BHE with a bigger drillhole diameter has a better thermal performance, despite the fact that the grouts have a lower thermal conductivity as compared to the ground. The investigated daily performance between 180 mm and 121 mm diameter is 6.7%, whereas this value drops to 2.16% between 165 mm and 121 mm diameter on a typical summer day. In addition, the difference of cooling performance among the three drillhole diameters depends heavily on the subterranean borehole temperature. A lower the subterranean borehole temperature yields a higher BHE cooling performance with a bigger drillhole diameter. In the long run, the amount of seasonal thermal exchange for the BHE of 165 mm and 180 mm diameter is 3.2% and 7.1% larger than that of 121 mm diameter, respectively.

NOMENCLATURE:

T temperature (°C)

R thermal resistance (W/m/K)

r borehole radius (m)

Q heat rate (kW)

W pumping rate (m^3)

H borehole length (m)

q thermal transfer rate (W/m)

E thermal load (kWh)

C specific thermal capacity (J/kg/K)

s water head (m)

KD hydraulic tansmissivity

Greek Symbols:

ρ density (g/cm^3)

λ thermal conductivity (W/m/K)

Subscripts:

i inlet flow

o outlet flow

f fluid

s ground surrounding

eff effective

b borehole

w water

ACKNOWLEDGEMENTS

This work is financed by Bayerisches Staatsministerium für Umwelt und Gesundheit. Sincere thanks are due to the Bayerische Forschung Stiftung (BFS) for providing a scholarship. The authors would also like to thank for the support provided by the Ochs Company (Nuremberg, Germany). We acknowledge support by the Deutsche Forschungsgemeinschaft and Friedrich-Alexander-Universität Erlangen-Nürnberg within the Open Access Publishing funding programme.

REFERENCES

1. Michopoulos, A.; Bozis, D.; Kikidis, P.; Papakostas, K.; Kyriakis, N.A. Three-years operation experience of a ground source heat pump system in Northern Greece. Energy Build. **2007**, 39, 328–334.
2. Yang, H.; Cui, P.; Fang, Z.H. Vertical-borehole ground-coupled heat pumps: A review of models and systems. Appl. Energy **2010**, 87, 16–27.
3. Bose, J.; Smith, M.D.; Spiter, J.D. Advances in Ground Source Heat Pump Systems—An International Overview. In Proceedings of the 7th International Energy Agency Heat Pump Conference, Beijing, China, 19–22 May 2002; pp. 313–324.
4. Gehlin, S. Thermal Response Test: In Situ Measurements of Thermal Properties in Hard Rock. Master's Thesis, Lulea University of Technology, Lulea, Sweden, 1998.
5. Teza, G.; Galgaro, A.; De, C.M. Long-term performance of an irregular shaped borehole heat exchanger system: Analysis of real pattern and regular grid approximation. Geothermics **2012**, 45–56.
6. Hwang, Y.J.; Lee, J.K.; Jeong, Y.M.; Koo, K.M.; Lee, D.H.; Kim, I.K.; Kim, I.K.; Jin, S.W.; Kim, S.H. Cooling performance of a vertical ground-coupled heat pump system installed in a school building. Renew. Energy **2009**, 34, 578–582.
7. Smith, M.D.; Perry, R.L. Borehole grouting: Field studies and thermal performance testing. ASHRAE Trans. **1999**, 105, 451–457.
8. Borinaga-Treviño, R.; Pascual-Muñoz, P.; Castro-Fresno, D.; Del Coz-Díaz, J.J. Study of different grouting materials used in vertical geothermal closed loop heat exchangers. Appl. Therm. Eng. **2012**, 50, 159–167.

9. Philippacopoulos, A.J.; Berndt, M.L. Influence of debonding in ground heat exchangers used with geothermal heat pumps. Geothermics **2001**, 30, 527–45.

10. Pahud, D.; Matthey, B. Comparison of the thermal performance of double U-pipe borehole heat exchangers measured in situ. Energy Build. **2001**, 33, 503–507.

11. Borinaga-Treviño, R.; Pascual-Muñoz, P.; Castro-Fresno, D.; Blanco-Fernández, E. Borehole thermal response and thermal resistance of four different grouting materials measured with a TRT. Appl. Therm. Eng. **2013**, 53, 13–20.

12. Lee, C.; Park, M.; Min, S.; Kang, S.; Sohn, B.; Choi, H. Comparison of effective thermal conductivity in closed-loop vertical ground heat exchangers. Appl. Therm. Eng. **2011**, 31, 3669–3676.

13. Desmedt, J.; Van Bael, J.; Hoes, H.; Robeyn, N. Experimental performance of borehole heat exchangers and grouting materials for ground source heat pumps. Inter. J. Energy Res. **2012**, 36, 1238–1246.

14. Zeng, H.; Diao, N.R.; Fang, Z.H. Heat transfer analysis of boreholes in vertical ground heat exchangers. Int. J. Heat Mass Transf. **2003**, 46, 4467–4481.

15. Shu, H.; Duanmu, L.; Hua, R. Analysis of Selection of Single or Double U-bend Pipes in a Ground Source Heat Pump System; Energy Systems Laboratory, Texas A&M University: College Station, TX, USA, 2006.

16. Choi, J.C.; Park, J.; Lee, S.R. Numerical evaluation of the effects of groundwater flow on borehole heat exchanger arrays. Renew. Energy **2013**, 52, 230–240.

17. Wang, H.J.; Qi, C.Y.; Du, H.P.; Gu, J.H. Thermal performance of borehole heat exchanger under groundwater flow: A case study from Baoding. Energy Build. **2009**, 41, 395–401.

18. Zanchini, E.; Lazzari, S.; Priarone, A. Long-term performance of large borehole heat exchanger fields with unbalanced

seasonal loads and groundwater flow. Energy **2012**, 38, 66–77.

19. Lazzari, S.; Priarone, A.; Zanchini, E. Long-term performance of BHE (Borehole Heat Exchanger) fields with negligible groundwater movement. Energy **2010**, 35, 4966–4974.

20. Lamarche, L.; Beauchamp, B. New solutions for the short-time analysis of geothermal vertical boreholes. Int. J. Heat Mass Transfer. **2007**, 50, 1408–1419.

21. Bernier, M.A. Long-term ground-temperature changes in geo-exchange systems. ASHRAE Trans. **2008**, 114, 342–350.

22. Sanner, B.; Hellström, G.; Spitler, J.; Gehlin, S. Thermal Response Test—Current Status and World-Wide Application. In Proceedings of the World Geothermal Congress, Antalya, Turkey, 24–25 April 2005.

23. Pröll, M. Durchführung von 3 Thermal Response Tests in Nürnberg; Technical Report for Bayerisches Zentrum für Angewandte Energieforschung: Nuremberg, Germany, 2012.

24. Hellström, G. Ground Heat Storage, Thermal Analysis of Duct Storage Systems. Ph.D. Thesis, University of Lund, Lund, Sweden, 1991.

25. Blomberg, T.; Claesson, J.; Eskilson, P.; Hellström, G.; Sanner, B. EED 3.0 User's Manual. Available online: http://www.buildingphysics.com/manuals/EED3.pdf (accessed on 27 December 2012).

26. Graphic Data Manager, RSG40 Memograph M, Energy Option. Available online: http://www.de.endress.com/#product/RSG40?open&tab=documents&filter=010.030 (accessed on 27 September 2012).

10
CHAPTER

EXPERIMENTAL STUDY ON FROST HEIGHT OF ROUND PLATE FIN-TUBE HEAT EXCHANGERS FOR MOBILE HEAT PUMPS

Moo-Yeon Lee [1], Yongchan Kim [2] and Dong-Yeon Lee [3,*]

[1] *Department of Mechanical Engineering, Dong-A University, 37 Nakdong-Daero 550 beon-gil saha-gu, Busan, Korea*

[2] *Department of Mechanical Engineering, Korea University, Anam-Dong, Sungbuk-Ku, Seoul 136-713, Korea*

[3] *School of Mechanical Engineering, Yeungnam University, 280 Daehak-Ro, Gyeongsan, Gyeongbuk 712-749, Korea*

ABSTRACT

The objective of this study was to provide experimental data that could be used to predict frost growth and frost performance of a round plate fin-tube heat exchanger for low temperature heat pumps used in zero emission vehicles under cold weather conditions. In this study, round plate fin-tube heat exchangers were tested with variation of the fin space, air flow rate, relative humidity, and inlet air temperature. Frost height was measured and considered with the boundary layer interruption between fins. Frost height for 8.0 mm of fin space was increased by approximately 91.9% with an increase of relative humidity from 50.0% to 80.0%. The growth rate of frost height at 1.2 m^3/min was observed to be 13.0% greater than that at 0.8 m^3/min. Finally, the variation of the blockage ratio with fin space would be an important reference for designing advanced heat exchangers that operate under cold weather conditions.

Keywords: frost; heat exchanger; heat pump; height; round plate fin

1. INTRODUCTION

The formation of frost is a general physical phenomenon that occurs under cold weather conditions in air to air and air to refrigerant heat exchangers, such as the evaporator of an air conditioning system, heat exchangers in a refrigeration system, heat pumps, in the storage of cryogenic liquids, and on the Earth's surface in winter. Accordingly, frost formation and growth is an important problem that is often encountered in refrigeration systems, especially in domestic and commercial refrigerators and freezers, heat pumps used in cold weather conditions, and recently mobile heat pumps for vehicles. As humid air passes over the cold fin of the heat exchanger, which is below the freezing point, water condenses onto the cold plate surface. Moisture molecules migrate to frosted surface and the void of frost layer leads to frost growth and densification of frost layer. Frost layer acts as an insulating layer between the fins and the air. In addition, frost blocks the air path, causing an increased pressure drop and a decrease in the air flow rate. Usually frost is removed periodically to improve the efficiency of operation. This normally involves heating the surface to melt frost, a costly operation since energy is required to melt frost and the heat exchanger is out of operation during the process, although several researchers suggested the non-costly defrosting techniques for heat pump systems at a laboratory stage as mentioned in Huang et al. [1] and Byrne et al. [2].

Hence frost properties have to be exactly predicted in order to build a numerical model of frost growth. Density and thermal conductivity, which affect the mass and heat transfer rate under cold weather conditions, are interrelated and are known to be functions of the air velocity, temperature, surface temperature, and air humidity ratio. Numerous experimental and theoretical investigations on frosting have been performed over a long period of time because the prediction of frost growth with time in the heat exchanger is directly related with frost heat transfer rate, the amount of frost, and the defrosting time. Jones and Parker [3] studied frost formation with varying environmental parameters. They developed a model for prediction of frost growth. The trends predicted by their model were in good agreement with the test data. However, their models did not consider molecular diffusion in frost layer. Brian et al. [4] developed a frost growth model which considered frost layer as a porous substance. Neiderer [5] studied frosting and defrosting effects on heat transfer coils. Kondepudi and O'Neal [6] studied on the effect of frost growth on the performance of louvered finned tube heat exchangers. O'Neal et al. [7] studied on the impact of a hydrophobic coating on frost buildup and the defrost performance of a heat pump heat exchanger. Rite and Crawford [8] studied experimentally the effects of various parameters on the rate of frost formation on the domestic refrigerator-freezer finned tube heat exchanger coil with 5.1 mm of fin space. Hayashi [9] found that the amount of frosting gets larger at the entrance of an air flow, in agreement with the general results of other papers. Schneider [10] simulated frost height by balancing the heat transfer rate between vapor diffusivity on frost surface and thermal conductivity between frost layers of a parallel plate heat exchanger. Sanders [11] studied the influence of frost formation and defrosting on the performance of air coolers. Sommers and Jacobi [12,13] studied the air-side heat transfer enhancement of a refrigerator evaporator that used the winglet vortex generation technique under both non-frosting and frosting conditions. Xia et al. [14] studied frost, defrost, and refrost and its impact on the air-side thermal-hydraulic performance of louvered-fin, flat-tube heat exchangers. Xie et al. [15] developed multiple correlations on the air-side heat transfer and friction characteristics of fin-tube heat exchangers by parametric study. Shao et al. [16] showed the comparison of heat pump performance using fin and tube and microchannel heat exchangers under frosting conditions. Hong et al. [17] studied frost growth of the louvered fins of a micro channel heat exchanger. Moallem et al. [18] reported experimental measurements on frosting performance of the micro channel heat exchangers for heat pump system. These researches were mostly concerned with general heat exchangers in refrigeration systems and the HVAC system in simple geometries.

However, the majority of papers on frosting in air conditioning systems, heat pumps, and refrigeration systems, which have more complex geometries such as fin-and-tube heat exchangers, are quite limited due to the large number of variables, the difficulty in quantitative analysis of frost, and the thermodynamic properties of humid air. The round plate fin-tube heat exchanger having large fin pitch ranging from 4.0 to 12.0 mm is considered in this study, which is much larger than the fin pitch of heat exchangers used for the residential and commercial heat pump systems (normally, fin pitch ranging from 1.4 to 2.0 mm). This is because frosting problem is very important in adoption of the heat exchanger used in the heat pump system for zero emission vehicles such as electric vehicles, fuel cell electric vehicles, hybrid electric vehicles, etc. So, the round plate fin-tube heat exchangers with large fin pitches might be used for zero emission. The reason is that the use of heat pumps instead of a PTC (positive temperature coefficient) heater for heating in a cabin under cold weather conditions has been studied for the longer driving ranges of the above-mentioned vehicles [19]. In addition, there could be an easy defrosting water disposal option through an outdoor installation of the outdoor heat exchanger for mobile heat pumps because of the round fin configurations. In this respect, new and precise studies on frosting behavior of round plate fin tube heat exchangers under cold weather conditions are necessary. The objective of this study is to investigate frost behavior and frost growth of a heat exchanger with a round plate fin configuration and large fin pitch under cold weather conditions. Finally, these experimental data could be used for designing an effective heat exchanger of a heat pump system to be used in zero emission vehicles.

2. EXPERIMENTAL SETUP

2.1. Test Specifications

A schematic of the experimental setup used to measure frost height during frost growth on the tested heat exchangers is shown in Figure 1. The test equipments were installed in a psychrometric chamber to provide a pre-controlled ambient temperature. The psychrometric chamber was maintained at 5.0 °C using an air-handling unit including a cooling coil, a heating coil, and a humidifier. The air temperature entering the test section was adjusted by using an electric resistance heater installed at the entrance of the flow chamber. The humidity was maintained at a set point by adjusting the power input to an ultrasonic humidifier. For measurement convenience, an ethylene glycol-water mixture was utilized as a refrigerant which passed through aluminum tubes. The temperature of the ethylene glycol-water mixture was constantly controlled properly by a constant temperature bath including a chiller and an electric heater. Air temperatures of test coils at the inlet and outlet were measured by arithmetically averaging the values obtained with twelve thermocouples installed in a uniform grid at each location. Frost layer surface temperature during frost growth was measured using an infrared thermometer with an accuracy of ±0.5 °C. Frost height was measured using a high performance digital cam-coder and a micro-scale measurement device with an accuracy of ±0.001 mm. The thermocouples were calibrated to an accuracy of ±0.1 °C. The absolute humidity was measured using a chilled mirror dew point sensor with an accuracy of ±0.2 °C. The air flow rate was measured using a nozzle in an open wind tunnel to an estimated accuracy of ±2.0%. The pressure drop across the tested heat exchanger was measured using a differential pressure transducer with an accuracy of ±1.0%. The refrigerant flow rate was measured using a Coriolis effect flow meter with an uncertainty of ±0.2% of the reading.

Figure 2 shows the schematic diagram of the tested round fin tube heat exchanger. Table 1 shows the test conditions and specifications in this experiment. Test conditions are considering the real operating conditions of a heat exchanger for mobile heat pumps under winter season used in this study. Especially, inlet air temperatures were chosen by reflecting effects of the waste heat from the electric devices of zero emission vehicles under winter season. The round plate fin-tube heat exchangers were tested by varying the air flow rate, relative humidity, fin space, and inlet air temperature. The tested heat exchangers were designed to allow slip space in the front end module (FEM) in the vehicles. The FEM consisted of outdoor heat exchangers (condenser or evaporator), radiators for electric devices and stacks, and a cooling fan used at the front of zero emission passenger vehicles. The diameters of the tube and the round fin were 8.2 mm with the tube thickness of 0.5 mm thickness and 23.0 mm with the fin thickness of 0.15 mm, respectively. Fins were mechanically expanded and coupled with tubes by a fin-tube press machine. Fin spaces were varied from 4.0 mm to 12.0 mm considered as heat exchanger usage in the mobile heat pump for zero emission vehicles. Inlet air temperature, inlet refrigerant temperature, and refrigerant mass flux were kept constant in the experiments. In order to observe frost growth and height, a top-of-the-range cam-coder was used to measure frost height on the round plate fin at a real scale. Frost height was measured by micrometer with an accuracy of 0.001 mm. All test conditions and specifications were selected to reflect the use of the heat exchanger for a mobile heat pump in cold weather conditions. Especially, in this study, the temperature of the ethylene glycol-water mixture was chosen for reflecting the lowest evaporating temperature, which could be operated in the mobile heat pump.

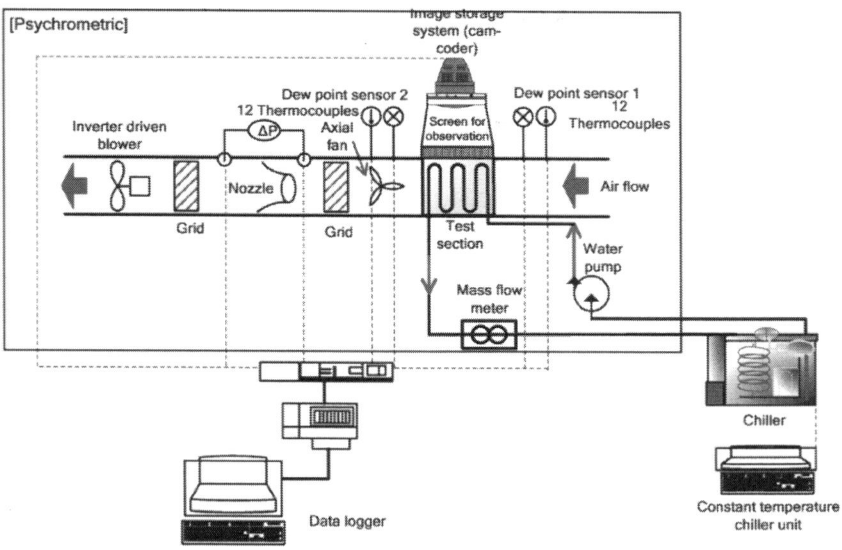

Figure 1. Schematic diagram of the test set-up.

2.2. Frost Height Measurement

Frost height was tested for varying air flow rate, relative humidity, and inlet air temperature. Frost height of the tested heat exchanger was measured at the middle fin in first tube of the round plate fin-tube heat exchanger having 2 rows and 2 columns in series because the deviation of frost height at all fins of first tube can be negligible. Namely, frost height measurement location was situated on a circle of a diameter that was half the diameter of the fin and frost height at the fin was a direct distance from center point of the fin surface to end point of frost (column). The measurement location of frost height is 5.75 mm from the center of the round plate fin as shown in Figure 2. Frost at each fin in the first tube grew as nearly uniform thickness because air flow through the fins was uniform. In addition, variations of the surface roughness of frost with time were not considered in this study because of the relatively large interval time. Frost growth patterns were observed using a high performance digital cam-coder at the middle fin of the tested heat exchanger. Frost height was measured every 10 minutes until 100 minutes. Effect of a surface coating on the defrosting water disposal of the tested round fin tube heat exchanger was not considered in this study. In addition, temperature of the ethylene glycol-water mixture was constantly controlled during the test. In this study, all measured frost height and growth were observed at the round plate fin. In addition, the frost pattern evolution was observed: it started by a crystal growth period and reached a frost layer growth period without reaching the frost layer full growth period as defined by Lee et al. [20].

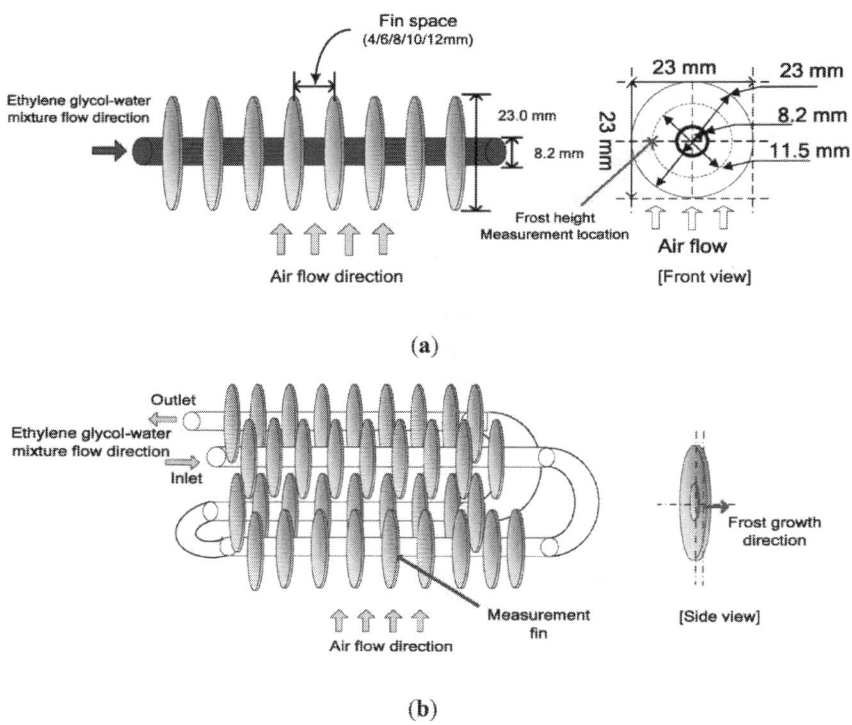

Figure 2. Specifications and configuration of the round plate fin-tube heat exchanger. (**a**) Round plate fin and tube specification; (**b**) Heat exchanger configuration and frost height measurement location.

Table 1. Test conditions and specifications.

Parameter	Value
Fin space (mm)	4.0/6.0/8.0/10.0/12.0
Fin diameter (mm)	23.0
Fin thickness (mm)	0.15
Tube outer diameter (mm)	8.2
Tube thickness (mm)	0.5
Tube length (mm)	650.0
Materials for a fin and a tube	Aluminum
Inlet air temperature (°C)	5.0/8.0/11.0
Inlet air relative humidity (%)	50/60/70/80
Airflow rate (m^3/min)	0.6/0.8/1.0/1.2
Ethylene glycol-water mixture inlet temperature (°C)	−15.0

3. RESULTS

3.1. Boundary Layer

Figure 3 shows the definition of the boundary layer of the round plate fins used in this study. The boundary layer thickness was determined by solving the continuity, momentum, and energy equation. Governing equations and boundary conditions can be found in Incropera and Dewitt [21]. The appropriate boundary layer conditions were expressed by Equation (1). Air boundary conditions are considered as free stream.

$$\begin{aligned} &u(x,0) = v(x,0) = 0 \quad \text{and} \quad u(x,\infty) = u_\infty \\ &T(x,0) = T_s \quad \text{and} \quad T(x,\infty) = T_\infty \end{aligned} \tag{1}$$

Figure 3(a) shows the schematic of the boundary layer between the round plate fins. Boundary layer interruption between the fins should be considered because it caused the heat transfer coefficient to significantly decrease. To enhance the heat transfer coefficient of the tested heat exchanger, the boundary layer interruption between the fins should be prevented and the developed boundary layer along the fins between the tubes should be broken as mentioned by Lee [22]. Also, the boundary layer thickness increases with the increase of the plate length, resulting in the decrease of the heat transfer coefficient as mentioned by Kim and Kim [23]. Leading edge of the boundary layer formation in the round plate fin was changed along the contour of the round plate fin as shown in Figure 3(b). The boundary layer length and thickness was changed with the contour shape of the round plate fin as given in Equation (2). $L_{x,crit}$, which is the distance after occurrence of the boundary layer interruption between the fins, in Equation (3) was the position of a fully developed boundary layer, or the distance from the leading edge at which transition begins. $L_{x,c}$ is the occurrence point of the boundary layer interruption between the fins.

$$L_x = x_o + (r - \sqrt{r^2 - y^2}) \tag{2}$$

$$L_{x,\mathrm{crit}} = L_{x,c} + (r - \sqrt{r^2 - y^2}) \tag{3}$$

3.2. Frost Growth

Figure 4(a) shows frost height as the fin space is varied from 4.0 mm to 12.0 mm at an inlet air temperature of 5.0 °C, relative humidity of 70.0%, air flow rate of 1.0 m^3/min and refrigerant temperature of −15.0 °C. Frost height for the 8.0 mm fin space showed a similar rate of increase during the first step of 10 minutes because boundary layer interruption theoretically did not occur at over 7.0 mm of fin space during the tests. As frost continued to grow over 25.0 minutes, frost growth rate in each fin space differed because of the boundary layer interruption by frost growth. Frost growth rate for the 8.0 mm fin space was lower compared with other fin spaces because the boundary layer interruption occurred at the beginning of frost formation. Theoretically, frost growth rate was affected by the fin space as mentioned by Mon and Gross [24]. However, frost height in the case of the larger fin space of 10.0 mm was less affected by the fin space because the boundary layer interruption was delayed. Figure 4(b) shows the blockage ratio with the variation of the fin space. Frost blockage ratio was defined as Equation (3), suggested by Lee et al [25]. Frost blockage ratio between the fins was defined as the rate of decrease of the air flow path due to the increase in frost height with time. Frost blockage ratio increased with decreasing fin space. Therefore, knowledge of frost blockage ratio would be an important reference for the design of an advanced heat exchanger to operate under cold weather conditions. Frost blockage ratio decreased on average by 26.1% with the rise of the fin space from 6.0 mm to 12.0 mm. Therefore the design of the fin space in a heat exchanger is a key parameter when considering the operation time under frosting conditions. In addition, the blocking of the air flow area at smaller fin pitches was more accelerated with frost growth.

$$\gamma = 2\frac{X_s}{F_f}\cdot 100\ (\%) \qquad (4)$$

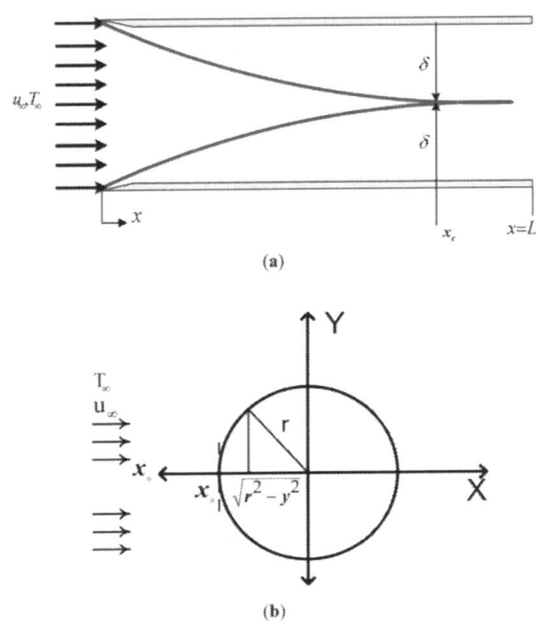

Figure 3. Boundary layer of the round plate fins. (**a**) Schematic of the boundary layer between the fins; (**b**) Configurations of the round plate fin.

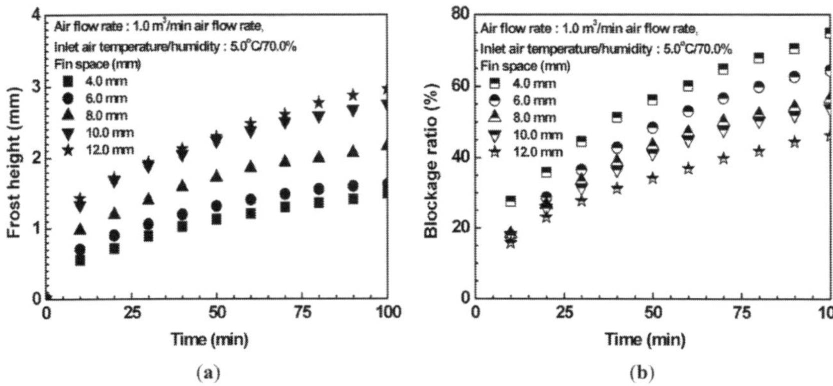

Figure 4. Frost height with variation of the fin space. (**a**) Frost height; (**b**) Blockage ratio.

Figure 5 shows frost height with the variation of the relative humidity of the inlet air for a fin space of 10.0 mm, inlet air temperature of 5.0 °C, air flow rate of 1.0 m^3/min and refrigerant temperature of −15.0 °C. Frost height at each relative humidity increased proportionally with time. Frost growth rate at a relative humidity ranging from 50.0% to 70.0% showed a similar increasing rate with time but frost growth rate at a relative humidity of 80.0% increased steeply for the first 40 minutes then grew uniformly. Frost height at 80.0% relative humidity was 91.9% greater than that at 50.0% relative humidity because humidity ratio at 80.0% relative humidity was 51.0% greater than that at 50.0% relative humidity.

Figure 6 shows frost height with the variations of the air flow rates from 0.8 m^3/min to 1.2 m^3/min for a fin space of 10.0 mm, inlet air temperature of 5.0 °C, relative humidity of 70.0% and refrigerant temperature of −15.0 °C. Generally, frost thickness increased slightly with air flow rate (or Reynolds number) because of rises in the heat and mass transfer rates with increased air flow rate on the fin surface. The reason is that both the mass deposition rate of water vapor and the density of frost layer increased with increasing Reynolds number as mentioned by Sahin [26]. The author reported

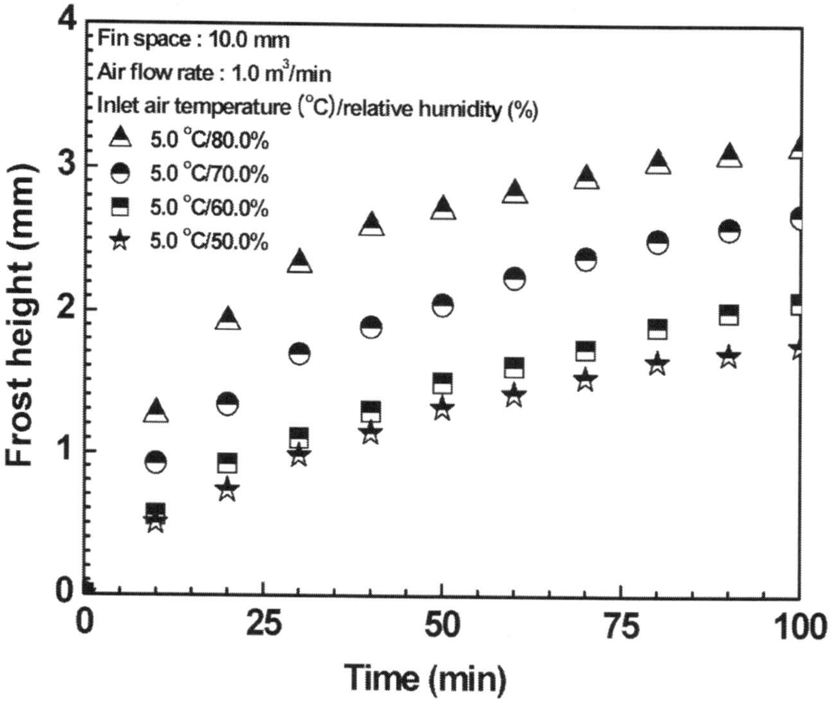

Figure 5. Frost height with the variations of the relative humidity.

that these two properties, almost offsetting each other, yielded an almost unchanged frost thickness. However, there is not yet fully elucidated as there are a number of conflicting data sets: some authors claim that a higher Reynolds number yields a higher frost layer, and some argue the opposite. Both maybe correct in view of the compensatory effect between both, since one of the properties (frost density or deposition rate of mass of water vapor) may be superior to the other under different conditions. Accordingly, in Figure 6, frost height increased with increasing air flow rate because of the increase of the heat and mass transfer rates, although the difference of the increasing rate of frost height was not important. Frost grew steeply until 1.5 mm of frost height at all air flow rates, but the increasing rate of frost height decreased because of the boundary layer interruptions between the fins with frost growth. The rate of increase of frost height at 1.2 m^3/min was observed to be

13.0% greater than that at 0.8 m^3/min. This is because the decreasing rates of the air flow rate with time at 1.2 m^3/min and 0.8 m^3/min were 25.0% and 15.8%, respectively. As a result, a compensatory effect between both, as mentioned above, appeared in the result in Figure 6.

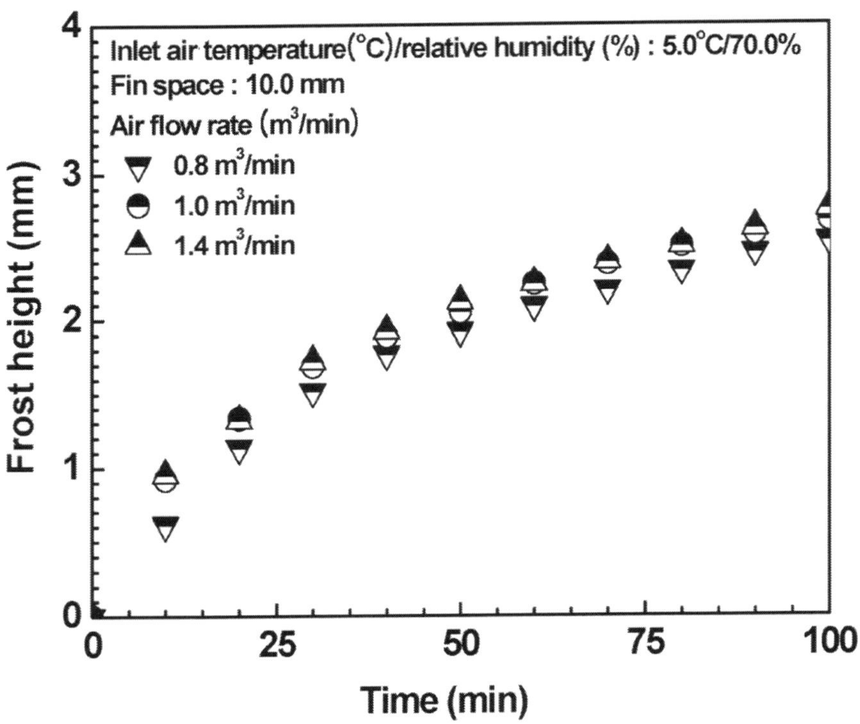

Figure 6. Frost height with the variations of the air flow rate.

Figure 7 shows frost height with variations of the inlet air temperature at fin space of 10.0 mm, refrigerant temperature of −15.0 °C, and air flow rate of 1.0 m^3/min. Frost height at an inlet air temperature of 11.0 °C was 38.5% greater than that at the inlet air temperature of 5.0 °C. However frost height at inlet air temperature of 11.0 °C was 4.5% greater than that of the inlet air temperature of 8.0 °C. This is because the humidity ratio increased with the increase of the inlet air temperature under same relative humidity but the increasing rate of humidity ratio decreased with the inlet air temperature increases. The increasing rate of the humidity ratio at inlet air temperatures from 5.0 °C to 11.0 °C and from 8.0 °C to 11.0 °C were 125.0% and 50.0%, respectively. Finally, major parameters affected by frost height were the fin space, relative humidity, inlet air temperature, and air flow rate in a row. These trends appeared in the results of Lee et al. [27]. Moreover, the fin space was critically related to the boundary layer interruption with frost growth and air flow rate reduction with time. In order to develop a more precise frost growth model, the boundary layer interruption between fins should be considered to predict frost density with frost growth.

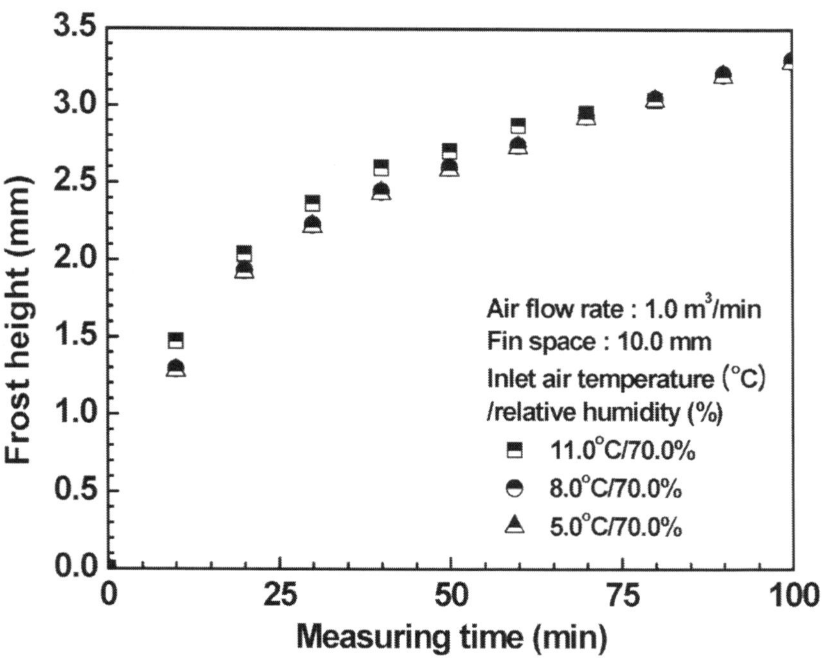

Figure 7. Frost height with the variations of the inlet air temperature.

4. SUMMARY AND CONCLUSION

Frost height and frost growth rate on the round plate fin-tube heat exchanger, which could be used as the heat exchanger of a heat pump in a zero emission vehicle, as well as in a refrigeration system or an HVAC system in various industries, were experimentally considered. Frost height was affected by the relative humidity, inlet air temperature, and air flow rate. The fin pitch was critically related to boundary layer interruption with frost growth. The variation of the blockage ratio with fin space would be an important reference for designing advanced heat exchangers that operate under cold weather conditions.

(i) Frost height for 8.0 mm of fin space was constantly increased during the first 10 minutes because the boundary layer interruption theoretically did not occur for a fin space during the tests.

(ii) Frost height increased by approximately 91.9% with an increase of relative humidity from 50.0% to 80.0%. Frost height increased by approximately 13.0% with an increase of the air flow rate from 0.8 m^3/min to 1.2 m^3/min.

(iii) Frost height at 70.0% relative humidity increased by 38.5% with the increase of the inlet air temperature from 5.0 °C to 11.0 °C. However, frost height at 11.0 °C inlet air temperature increased to 4.5% greater than that of the 8.0 °C inlet air temperature.

NOMENCLATURE

F_f Fin space (mm)

L Appropriate characteristic length (mm)

T Temperature (°C)

X_s Frost layer height (mm)

Greek symbols

δ Boundary layer thickness

r Radius (mm)

γ Blockage ratio (%)

u Velocity component in x direction (mm/s)

v Velocity component in y direction (mm/s)

x X-axis

y Y-axis

Subscripts

c current

crit critical

s surface

∞ free stream

ACKNOWLEDGMENTS

This research was supported by the Yeungnam University research grants in 2011.

REFERENCES

1. Huang, D.; Li, Q.; Yuan, X. Comparison between hot-gas bypass defrosting and reverse-cycle defrosting methods on an air-to-water heat pump. Appl. Energy **2009**, 86, 1697–1703.

2. Byrne, P.; Miriel, J.; Lénat, Y. Experimental study of an air-source heat pump for simultaneous heating and cooling, Part 2: dynamic behaviour and two-phase thermosiphon defrosting technique. Appl. Energy **2011**, 88, 3072–3078.

3. Jones, B.W.; Parker, J.D. Frost formation with varying environmental parameters. J. Heat Transfer **1975**, 97, 255–259.

4. Brian, P.I.; Reid, R.C.; Shah, Y.T. Frost deposit on cold surfaces. Ind. Eng. Chem. Environ. **1969**, 9, 375–380.

5. Niederer, D.H. Frosting and defrosting effects on coil heat transfer. ASHRAE Trans. **1976**, 82, 467–473.

6. Kondepudi, S.N.; O'Neal, D.L. The effect of frost growth on the performance of louvered finned tube heat exchangers. ASHRAE Trans. **1989**, 93, 258–274.

7. O'Neal, D.L.; Bryant, J.A.; Parker, B. Impact of hydrophobic coating on frost buildup and defrost performance of a heat pump heat exchanger. In Proceedings of the 45th Oji International Seminar, Tomakomai, Japan, September 1997; pp. 181–187.

8. Rite, R.W.; Crawford, R.R. The effect of frost growth on the performance of domestic refrigerator-freezer finned tube heat exchanger coils. ASHRAE Trans. **1991**, 97, 428–437.

9. Hayashi, Y.; Aoki, A.; Adachi, S.; Hori, K. Study of frost properties correlating with frost formation types. J. Heat Transfer **1977**, 99, 239–245.

10. Schneider, H.W. Equation of growth rate of frost formation on the performance of a parallel plate heat exchanger. Int. J. Heat Mass Transfer **1978**, 21, 1019–1024.

11. Sanders, C.T. The influence of frost formation and defrosting on the performance of air coolers. Ph.D. Thesis, Delft Technical University, Delft, The Netherlands, October 1974.

12. Sommers, A.D.; Jacobi, A.M. Air-side heat transfer enhancement of a refrigerator evaporator using vortex generation. Int. J. Refrig. **2005**, 28, 1006–1017.

13. Sommers, A.D.; Jacobi, A.M. Heat transfer enhancement by winglet-type vortex generator array in compact fin-and-tube heat exchangers. Int. J. Refrig. **2008**, 31, 87–97.

14. Xia, Y.; Zhong, Y.; Hrnjak, P.S.; Jacobi, A.M. Frost, defrost, and refrost and its impact on the air-side thermal-hydraulic performance of louvered-fin, flat-tube heat exchangers. Int. J. Refrig. **2006**, 29, 1066–1079.

15. Xie, G.; Wang, Q.; Sunden, B. Parametric study and multiple correlations on air-side heat transfer and friction characteristics of fin-and-tube heat exchangers with large number of large-diameter tube rows. Appl. Therm. Eng. **2009**, 29, 1–16.

16. Shao, L.L.; Yang, L.; Zhang, C.L. Comparison of heat pump performance using fin-and-tube and microchannel heat exchangers under frost conditions. Appl. Energy **2010**, 87, 1187–1197.

17. Hong, T.; Moallem, E.; Cremaschi, L.; Fisher, D.E. Measurements of frost growth on louvered folded fins of micro-channel heat exchangers Part 1: Experimental methodology. ASHRAE Trans. **2012**, 118, 1–7.

18. Moallem, E.; Cremaschi, L.; Fisher, D.E.; Padhmanabhan, S. Experimental measurements of the surface coating and water retention effects on frosting performance of micro-channel heat exchangers for heat pump systems. Exp. Therm. Fluid Sci. J. **2012**, 39, 176–186.

19. Cho, C.W.; Lee, H.S.; Won, J.P.; Lee, M.Y. Measurement and evaluation of heating performance of heat pump systems using wasted heat from electric devices for an electric bus. Energies **2012**, 5, 658–669.

20. Lee, M.Y.; Kang, T.H.; Joo, Y.J.; Kim, Y.C. Heat transfer characteristics of spirally-coiled circular fin-tube heat exchanger under frosting conditions. Int. J. Refrig. **2011**, 34, 328–336.

21. Incropera, F.P.; Dewitt, D.P. Fundamentals of Heat and Mass Transfer, 4th ed.; John Wiley & Sons: Hoboken, NJ, USA, 2002.

22. Lee, M.Y. Heat and mass transfer characteristics of a spirally-coiled circular fin tube heat exchanger. Ph.D. Thesis, Korea University, Seoul, Korea, February 2010.

23. Kim, Y.H.; Kim, Y. Heat transfer characteristics of flat plate finned-tube heat exchangers with large fin pitch. Int. J. Refrig. **2005**, 28, 851–858.

24. Mon, M.S. Numerical investigation of air-side heat transfer and pressure drop in circular finned-tube heat exchangers. Ph.D. Thesis, Technische Universität Bergakademie Freiberg, Freiberg, Germany, February 2003.

25. Lee, K.S.; Lee, T.H.; Kim, W.S. An experimental study on the effects of design factors for the performance of fin-tube heat exchanger under cold weather conditions. Trans. KSME **1995**, 19, 2657–2666.

26. Sahin, A.Z. An experimental study on the initiation and growth of frost formation on a horizontal plate. Exp. Heat Transfer **1994**, 7, 101–119.

27. Lee, S.H.; L, M.Y.; Won, Y.J.; Kim, Y. Frost growth characteristics of spirally-coiled circular fin-tube heat exchangers. Int. J. Refrig. **2012**. submitted.

11
CHAPTER
TESTING OF A LOW-COST LOOP HEAT PIPE DESIGN

Cosimo Buffone

Microgravity Research Centre, Universitè Libre de Bruxelles, Bruxelles, Belgium

ABSTRACT

This paper presents and describes the test campaign of a low-cost Loop Heat Pipes (LHP) design. LHP have been around for many decades now. Their potential as passive heat transfer devices has been widely demonstrated in numerous both groundand space-based applications. One of the major disadvantages of LHP is their inherent high manufacturing cost; this is the main factor why LHP are still confined to niche/high end applications such as thermal management of spacecrafts. This paper proposes to use an alternative manufacturing design for the LHP evaporator, which is the main contributor to the overall LHP cost. Preliminary thermal results are also reported and briefly explained. Future work is needed to confirm the promising results discussed in this paper and address fully other issues such as tolerance of this LHP design to vibrations and accelerations typical of space missions.

Keywords: Loop Heat Pipe; Low Cost Evaporator

1. INTRODUCTION

Loop Heat Pipes (LHP) were firstly conceived in Russia in the mid 70's to address some of the heat transfer issues present on board spacecrafts [1] -[7] . One of the main issues in spacecrafts such as satellites is their long-life span, typically of the order of many years (stretching to over 2 - 3 decades as well). For this reason, passive solutions are generally preferred to active ones because of the inherent reliability and life issues problems associated with active components such as pumps.

LHP are basically completely passive heat transfer devices which are able to transfer heat between two (or more) remote locations which are kept at different temperatures [8] -[12] . LHP have been around for many decades and have been successfully tested in numerous applications, both ground and space ones. LHP are suitable for low (>20 W) and high (>12 kW) heat loads and can operate at low (−160°C) and high (>200°C) temperatures. LHP are not so much different from the basic Heat Pipe (HP) concept, and therefore as Heat Pipes they are very versatile. However, LHP are far superior to standard HP because the wick is only present in the evaporator section (which is typically a very small proportion of the overall LHP length), and they have separate liquid and vapour lines connecting the evaporator and the condenser. These factors allow the use of very fine powders to produce high-performing sintered wicks (typically Nickel or Titanium), and also remove the entrainment limitation due to counter-flowing liquid and vapour in intimate contact as in standard HP designs.

LHP design has changed very little over so many years. In particular, the evaporator layout is basically the same as it was invented in the late 70's. One of the main reasons why this fact has stood well the test of time is that LHP are very difficult heat transfer devices to predict. In particular, the evaporator is a very complicated part to predict: the primary wick is an intricate porous media inside which two-phase with phase change takes place; the complex network of vapour removal grooves on the primary wick outside surface and the associated heat and mass transfer inside the grooves makes it difficult to predict the hydro-thermal dynamics; the complex thermal coupling of the evaporator with the compensation chamber (liquid reservoir) makes the thermal control of LHP quite challenging. Several detailed studies have described different modelling strategies to predict LHP performance [13] -[15] .

Almost in parallel and independently from the invention of LHP, the Americans at NASA had conceived the Capillary Pumped Loop (CPL) [16] . CPL is very similar to LHP and their operational principle is fundamentally the same. CPL has the liquid reservoir remotely located from the evaporator. This is a fundamental difference as it allows the evaporator to be very small and fit on confined spaces where there are packaging issues. The remoteness of the liquid reservoir also allows the adoption of multiple evaporator designs as one single liquid reservoir can feed all the evaporators. The remoteness of the liquid reservoir allows for independent temperature control of the CPL by setting up the liquid reservoir temperature; that is why CPLs are referred as loops with cold-biased reservoir. This fact gives CPL a very fine temperature tuning capability which is by far superior to LHP designs. Typically CPL temperature can be controlled within ±1°C accuracy. However, CPL has a major issue with the evaporator priming at start-up. In fact, because there is no mechanical connection between the evaporator and liquid reservoir, it is likely that most of the liquid is inside the liquid reservoir instead of the evaporator; therefore, at start-up one must ensure that the CPL evaporator is primed before heat is applied to it. The evaporator priming is a very long and tedious procedure (can take between 30 min and a few hours). For this reason CPLs are typically not preferred to LHP; and only when LHP cannot be used because of their limitations, CPL should be chosen instead.

Several studies have also been published on miniature LHP [17] -[19] and high performance wicks [20] [21] in which it has been proven that LHP can easily been scaled down maintaining its superior thermal performance and wicks can be designed to improve performance of the evaporator.

One of the major problems with LHP is the manufacturing and assembling cost of the evaporator. The typical manufacturing steps of a LHP evaporator are: 1) powder sintering to produce the primary wick; 2) creation of the network of grooves on the wick outside surface; 3) insertion of the wick inside a heated metal jacket; 4) the integration of the compensation chamber with the evaporator by using the secondary wick and at times the bayonet. The creation of the network of grooves and the insertion of the evaporator inside the metal jacket are the most difficult steps which typically result in high scraps rates because of damage on the grooves.

Recent attempts to reduce the manufacturing cost of LHP have started appearing in the open literature [22] [23] . Much more is been done, but given the commercial potentials of these attempts, still few of these studies are published.

The present work suggests a different way of manufacturing LHP evaporators. It is proposed that only longitudinal grooves are created during the sintering process by the insertion of mandrels in intimate contact with the metal jacket. The metal powder is therefore sintered inside the metal jacket. This allows at once the complete removal of the groove creation and wick insertion inside the metal jacket after the sintering process. Therefore, manufacturing costs should be dramatically reduced; even the scrap rate should be significantly improved.

Copper powder with average pore size of 10 μm is used and 8 stainless steel mandrels (vapour removal channels) with ID of 2 mm are used. The jacket material is copper with 2.5 mm thickness. The mandrels are secured in contact with the internal surface of the jacket through the use of a specifically designed jig. The copper powder is introduced inside the jacket and subsequently pressed up to 250 MPa. The subsequent sintering process took place in a furnace with controlled atmosphere at a temperature of 850°C for 2 hours.

The manufacturing process of the wick is slightly different from the process described in a previous relevant work by Vasiliev et al. [24] where a “chemically active substance” is used for Nickel powder in order to avoid important shrinkage inside of the wick during sintering.

The drawback of such a design is that there are no circumferential grooves on the primary wick which may create circumferential temperature differences when heat is applied only to one side of the LHP evaporator. This penalty may be alleviated by the use of thicker metal jackets which should conduct more heat around the circumference. However, this results in larger thermal resistances of the evaporator. Much more work is clearly needed to see if this issue can be addressed.

2. PRELIMINARY THERMAL TEST RESULTS AND DISCUSSION

In this section some preliminary thermal tests performed on the low-cost LHP design with water as working fluid are presented and briefly discussed.

Figure 1 shows a not to scale schematic of the low-cost LHP with the thermocouple location. Dimensions of LHP parts are also given in millimetres in the same Figure 1. The evaporator, compensation chamber, condenser, liquid and vapour lines are also indicated in Figure 1. As can be seen from Figure 1, both the liquid and vapour lines have flexible parts (the tubes are basically coiled) that allow the rotation of the LHP condenser with respect to the evaporator.

A wire heater was wrapped around the evaporator sector and the total power delivered by the heater was controlled via a variac. Insulation was provided around the heater to limit the heat losses versus the ambient.

Figure 2 reports the LHP response at power ramp-up and down. Power (which is also reported in Figure 2 as the step-wise black curve) is varied between 0 and 100 W in successive steps after steady state readings were achieved with the thermocouples. Additionally, in Figure 2 for completeness also the ambient temperature (thermocouple number 8) was monitored during tests.

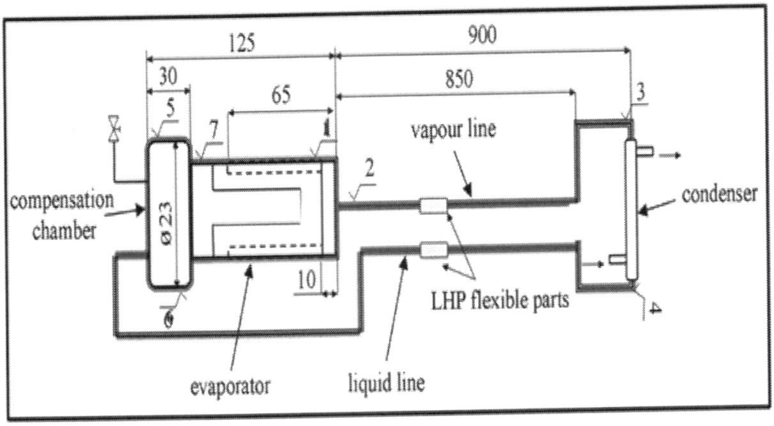

Figure 1. Schematic of loop heat pipe. 1 - 7 are K-Type thermocouples. All sizes are given in millimeters.

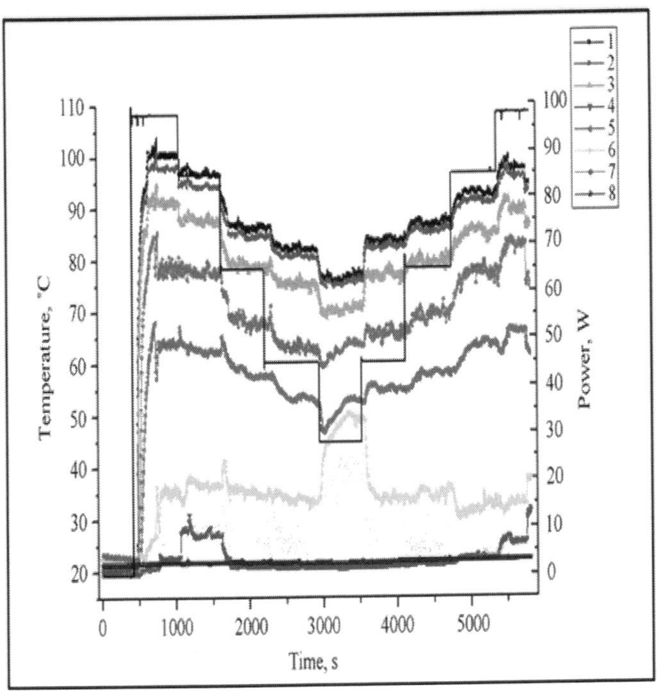

Figure 2. Power ramp-up and down. Power applied is given by the step-wise black curve. Thermocouple number 8 measures the ambient temperature.

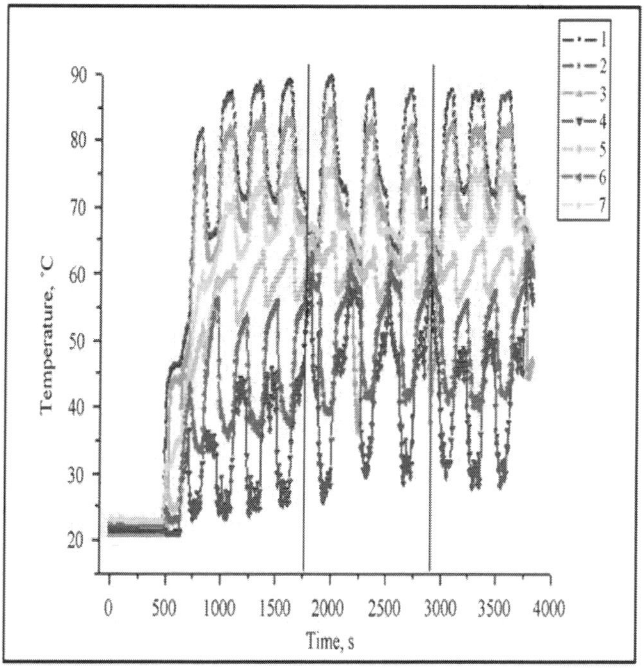

Figure 3. LHP response to power cycling between 20 and 60 W.

As can be seen from Figure 2 after the first power setting (100 W) was applied there is significant temperature raise in all but thermocouple 4, and a small but distinct temperature over-shoot which is necessary to displace all vapour bubbles inside the primary wick at start-up. The response of the LHP is around 200 s which is relatively large; this is believed to be due to the small diameter of the vapour line. Subsequently, the low-cost LHP design responds quite well at input power changes.

Power cycling tests have also been performed and they are presented in Figures 3 and 4. In Figure 3 power is changed between two set values (20 and 60 W). The LHP responsiveness is very good indeed over 1 hour of operation. In Figure 4 the power has been varied between two levels (45 and 90 W) for the first two cycles and for successive cycles the power has also been taken down to nil value. Again, the response of the LHP is good over around 40 minutes of operation.

In Figure 5 for a fixed power setting (85 W) the condenser temperature has been varied and the evaporator temperature monitored. It can be clearly seen that there is a quasi-linear relationship between the evaporator and condenser temperatures. There are two typical LHP operational modes, namely: constant conductance and variable conductance [10] . Inspection of Figure 5 suggests that the LHP is working in its constant conductance mode (where the equivalent thermal conductivity of the LHP is constant as in a normally functioning Heat Pipe).

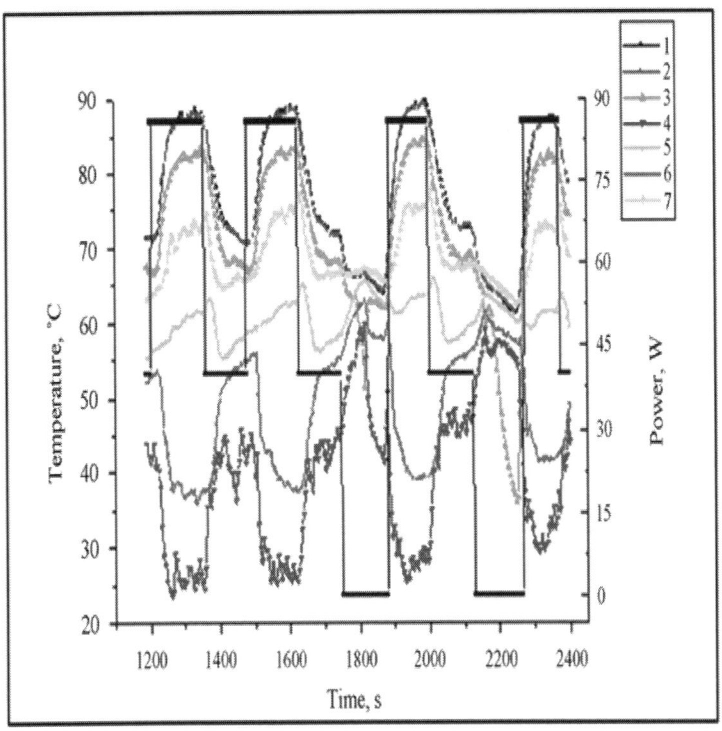

Figure 4. LHP response to power cycling between three power levels (0, 45 and 90 W).

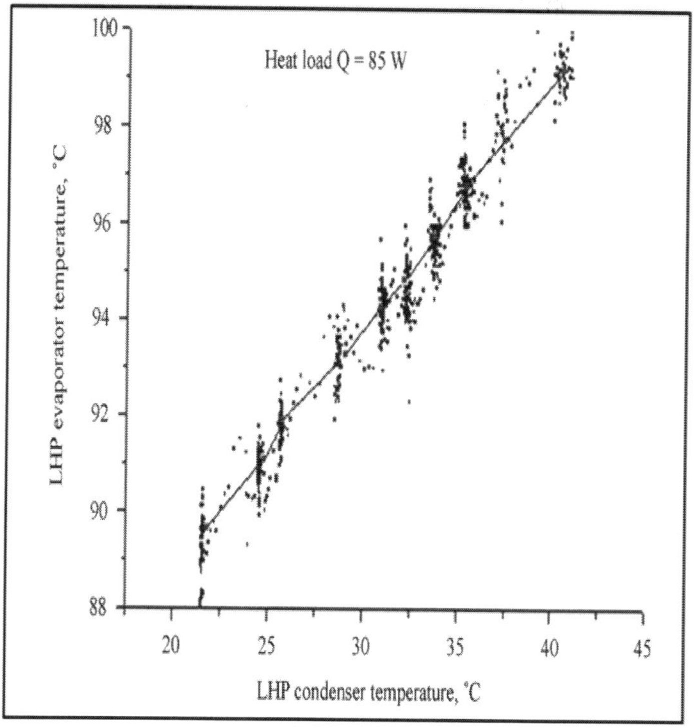

Figure 5. LHP evaporator temperature response to changes in condenser temperature for a fixed power level (85 W).

The thermal resistance defined as the temperature difference between LHP evaporator and condenser (estimated from Figure 5) and divided by the heat load is between 0.7 and 0.8°C/W. This range of thermal resistance is in line with the reported values for LHP in the literature [25] .

3. CONCLUSIONS

This paper presents some preliminary works carried out recently on a low-cost LHP design. The evaporator has been manufactured using only longitudinal grooves. The wick sintering has been carried out inside the metal jacket avoiding manufacturing and assembling issues typical of conventional LHP manufacturing. The only potential limitation of the proposed design is the absence of circumferential grooves which typically allow for vapour to be distributed uniformly around the circumference of the evaporator when heat is only applied to a sector of the circumference.

The preliminary thermal tests of the proposed low-cost LHP demonstrate good time response and fairly steady temperatures at power ramp-up and -down, power cycling and variation of condenser temperature.

More tests such as low power and high power endurance, and minimum power level at start-up should be performed to gain further confidence in this design as an alternative low-cost solution suitable for those applications which do not require very demanding thermal performance from a LHP.

ACKNOWLEDGEMENTS

The author would like to thank Dr. Leonid Vasiliev Jr. for performing most of the experimental tests.

REFERENCES

1. Maidanik, Y.F., Vershinin, S., Kholodov, V. and Dolggirev, J. (1985) Heat Transfer Apparatus. US Patent No. 4515209.
2. Delil, A.M., Maydanik, Yu.F. and Gernard, C. (2003) Development of Different Novel Loop Heat Pipes within the ISTC-1360 Project. SAE Paper No. 2003-01-2383.
3. Swanson, D.T. (2004) Thermal Control Technologies for Complex Spacecraft Systems. Proceedings of 13th International Heat Pipe Conference, Shanghai, 21-25 September 2004, China Astronautic Publishing House, Beijing.
4. Hoang, T.T., Cummings, S.M., Baldauff, R.W. and Brown, M.A. (2002) Advanced Loop Heat Pipes for Spacecraft Central Thermal Bus Concept. 12th International Heat Transfer Conference, Grenoble, August 2002.

5. Goncharov, A.K., Barantsevich, V.L. and Orlov, A.A. (2003) Experience of Development of Heat Pipes Applied in Russian Spacecrafts. Proceedings of Fifth Minsk International Seminar Heat Pipes, Heat Pumps, Refrigerators, Minsk, 8-11 September 2003.

6. Riehl, R.R. and Dutra, T. (2005) Development of an Experimental Loop Heat Pipe for Application in Future Space Missions. Applied Thermal Engineering, 25, 101-112.

7. Semenov, S.Y., Cho, W.L. and Jensen, S.M. (2006) High Conductance Loop Heat Pipes for Space Application. AIP Conference Proceedings, 813, 43.

8. Mishkinis, D. and Vasiliev, L. (2004) Heat Pipe Two-Phase Loop with Capillary Pump. The Update Journal of the Heat Exchanger Design Handbook, Begell House, Redding.

9. Vasiliev, L., Lossouarn, D., Romestant, C., Alexandre, A., Bertin, Y., Piatsiushyk, Y. and Romanenkov, V. (2009) Loop Heat Pipe for Cooling of High-Power Electronic Components. International Journal of Heat and Mass Transfer, 52, 301-308.

10. Maydanik, Y.F. (2005) Loop Heat Pipes. Applied Thermal Engineering, 25, 635-657.

11. Chernysheva, M.A., Vershinin, S.V. and Maydanik, Y.F. (2007) Operating Temperature and Distribution of a Working Fluid in LHP. International Journal of Heat and Mass Transfer, 50, 2704-2713.

12. Cheung, K., Hoang, T.T., Ku, J. and Kaya, T. (1998) Thermal Performance and Operational Characteristics of Loop Heat Pipe (NRL LHP). International Conference on Environmental Systems, Danvers, 13-16 July 1998.

13. Khrustalev, K. (2007) Modeling Transient Operation of Loop Heat Pipes Using Thermal Desktop. TFAWS, Cleveland, 10-13 September.

14. Parker, M.L. (2000) Modeling of Loop Heat Pipes with Applications to Spacecraft Thermal Control. Ph.D. Thesis, University of Pennsylvania, Pennsylvania.

15. Shukla, K.N. (2008) Thermo-Fluid Dynamics of Loop Heat Pipe Operation. International Communications in Heat and Mass Transfer, 35, 916-920.

16. Cullimore, B.A. (1993) Capillary Pumped Loop Application Guide. 23rd International Conference on Environmental Systems, Colorado Springs, 12-15 July 1993, SAE Paper No.932156.

17. Nagano, H., Nagai, H., Fukuyoshi, F. and Ogawa, H. (2008) Study of Thermal Performances of a Small Loop Heat Pipe. Journal of Thermal Sciences and Technology, 3, 355-367.

18. Maydanik, Y.F. (2004) Miniature Loop Heat Pipes. Proceedings of 13th International Heat Pipe Conference, Shanghai, 21-25 September 2004, China Astronautic Publishing House, Beijing.

19. Zhang, L., Xu, J. and Xu, H. (2013) Effect of Inventory on the Heat Performance of Copper-Water Loop Heat Pipe. Experimental Thermal and Fluid Science, 44, 875-882.

20. North, M.T., Sarraf, D.B., Rosenfeld, J.H., Maidanik, Y.F. and Vershinin, S. (1997) High Heat Flux Loop Heat Pipes. Sixth European Symposium on Space Environmental Control Systems, Noordwijk, 20-22 May 1997.

21. Rosenfeld, J.H., Sarraf, D.B., Khrustalev, D.K., Wellen, P.J. and North, M.T. (1999) Loop Heat-Pipe Evaporator with Bidisperse Wick Structures. NASA Tech Briefs.

22. Ku, J., Nagano, H., Perez, J. and Rogers, P. (2006) Thermal Performance of a Low-Cost Loop Heat Pipe. Annals of the Assembly for International Heat Transfer Conference, Begell House, Redding.

23. Vasiliev Jr., L.L. (2007) Advanced Loop Heat Pipe Evaporator with Ceramic Nanostructured Composite of Alumina,

Alumina-Silica Oxide as a Wick Structure. SAE Technical Paper 2007-01-3192.

24. Vasiliev Jr., L.L., Marengo, M., Ferrandi, C., Zinna, S. and Maziuk, V. (2009) Advanced Design of a "Low-Cost" Loop Heat Pipe. SAE Technical Paper 2009-01-2519.

25. Launay, S., Sartre, V. and Bonjour, J. (2007) Parametric Analysis of Loop Heat Pipe Operation: A Literature Review. International Journal of Thermal Sciences, 46, 621-636.

INDEX